Cholinesterase Genes: Multileveled Regulation

Monographs in Human Genetics

Vol. 13

Series Editor
Robert S. Sparkes, Los Angeles, Calif.

KARGER

Basel · München · Paris · London · New York · New Delhi · Bangkok · Singapore · Tokyo · Sydney

Cholinesterase Genes: Multileveled Regulation

Hermona Soreq
Department of Biological Chemistry, The Life Sciences Institute,
The Hebrew University of Jerusalem, Israel

Haim Zakut
Department of Obstetrics and Gynecology, The Edith Wolfson
Medical Center, Holon, The Sackler Faculty of Medicine of the
Tel Aviv University, Tel Aviv, Israel

17 figures and 4 tables, 1990

KARGER

Basel · München · Paris · London · New York · New Delhi · Bangkok · Singapore · Tokyo · Sydney

Monographs in Human Genetics

Library of Congress Cataloging-in-Publication Data
Soreq, H.
Cholinesterase genes: multileveled regulation / Hermona Soreq, Haim Zakut.
p. cm. — (Monographs in human genetics; vol. 13)
Includes bibliographical references.
1. Cholinesterase genes. 2. Cholinesterases – Synthesis – Regulation. 3. Genetic regulation.
I. Zakut, Haim, 1935–. II. Title. III. Series: Monographs in human genetics, v. 13.
[DNLM: 1. Cholinesterases – genetics. 2. Gene Expression Regulation, Enzymologic]
ISBN 3-8055-5137-1

Drug Dosage

The authors and the publisher have exerted every effort to ensure that drug selection and dosage set forth in this text are in accord with current recommendations and practice at the time of publication. However, in view of ongoing research, changes in government regulations, and the constant flow of information relating to drug therapy and drug reactions, the reader is urged to check the package insert for each drug for any change in indications and dosage and for added warnings and precautions. This is particularly important when the recommended agent is a new and/or infrequently employed drug.

Printed in Switzerland by Thür AG Offsetdruck, Pratteln
ISBN 3-8055-5137-1

Contents

III. Basic Research and Clinical Implications

Acknowledgements

We are grateful to Dr. J. Sadoff (Washington, D.C.) and to all members of our collaborative research group in Jerusalem and Tel Aviv for their practical and theoretical contributions to this work. Supported by Contract DAMD-17-87-7169 (to H.S.) and by the Research Fund at the Edith Wolfson Medical Center (to H.Z.). Special thanks to Mrs. Linda Milner.

I. Introduction: Multileveled Regulation of the Human Cholinesterase Genes and Their Protein Products

1. Overview and General Significance

Acetylcholinesterase (AChE) is an enzyme long noted for its essential role in the termination of neurotransmission at cholinergic synapses and neuromuscular junctions. As the target protein for a variety of neurotoxic compounds, including common agricultural insecticides and chemical warfare agents, research on this enzyme has profound implications for human health and well-being. A member of the highly polymorphic family of cholinesterases (ChE), proteins expressed in tissue-specific and differentially regulated fashions, AChE presents a powerful model for the basic scientific study of complex gene regulation and divergent pathways in protein biosynthesis.

Recent accumulated evidence, covered in a number of excellent reviews, has revealed the primary structure of human ChEs and the chromosomal localization of the CHE1 gene encoding the related enzyme butyrylcholinesterase (BuChE). Further analytical expression studies have demonstrated that the biochemical properties of human ChEs are mostly inherent to their primary amino acid sequence, and the discovery of ChE gene amplifications implicated AChE and BuChE in cholinergic influences on cell growth and proliferation. Abnormal expression of both ChEs, the amplification of the ACHE and CHE genes encoding these enzymes, and unusual BuChEmRNA transcripts have been variously associated with abnormal megakaryocytopoiesis, different leukemias, and malignant tumors of the brain and ovary. The suggestion has been put forth that sublethal exposure to organophosphorous (OP) poisons may, through its effect(s) on ChEs, create a selection pressure for ChE gene amplifications and, in turn, play a role in cancer etiology. This monograph presents the development and outcome of these basic research and clinically oriented studies, within the context of current research advances in the field of ChEs.

The family of ChEs has been the subject of intensive research for five decades, with a continuous increase in the number of studies being focused on this family of enzymes as well as on their scope and diversity [63, 333]. At the biochemical level, ChEs are highly polymorphic enzymes of broad substrate specificity [46]. At the biological level, it has been assumed for a long time that ChEs are involved in the termination of neurotransmission in cholinergic synapses and neuromuscular junctions.

According to the accepted classification of enzymes, ChEs belong to the B-type carboxylesterases on the basis of their sensitivity to inhibition by OP poisons [348]. ChEs are primarily classified according to their substrate specificity and by definition of their sensitivity to selective inhibitors into AChE (acetylcholine acetylhydrolase, EC 3.1.1.7) and BuChE (acylcholine acylhydrolase, EC 3.1.1.8) [225]. However, the complete scheme is much more complex. Further classifications of ChEs are based on their electrical charge, their level of hydrophobicity, the extent and mode of their interaction with membrane or extracellular structures and, last but not least, the multisubunit association of catalytic and noncatalytic 'tail' subunits composing together the biologically active ChE molecules [304, 305, 315].

Apart from the neuromuscular junction, ChEs are relatively concentrated in the brain. Therefore, ChEs in the mammalian brain have been studied extensively for many years. Colocalization of AChE with choline acetyltransferase in cerebral regions has indicated cholinergic and cholinoceptive properties to certain families of mammalian brain neurons [202]. Also, correspondence was shown between the dopamine islands and AChE patches in the developing striatum [139]. An immunohistochemical study of neuropeptides in cat striatum demonstrated that neurons, producing opiate peptides, are arranged to form mosaic patterns in register with the striosomal compartments visible by AChE staining [140]. Also, AChE immunoreactivity coexists with that of somatostatin in rat cerebral neuron cultures [90]. ChE staining in the mediodorsal nucleus of the thalamus and its connections in the developing primate brain were found to be transient [186], whereas the AChE activity persists to adulthood in the substantia nigra and caudate nucleus of the cat [143] and is subject to in vivo release under electrical stimulation [142] in parallel with dopamine release [71].

What is the role of ChEs in these different brain regions and which elements regulate the expression of CHE genes in the mammalian brain? These questions still remain unanswered. Various studies related AChE activities with neonatal learning [173] and other high functions. Other studies demonstrated hormonal regulation of AChE, for example by ecdysterone [67] in the mammalian brain or by β-ecdysone in *Drosophila* cells [66]. These studies, like the structural related ones, await more detailed molecular analysis.

Numerous studies over the years have indicated that the severe clinical symptoms resulting from intoxication by OP agents [52, 157, 171, 210, 43, 341] are caused by their very tight, irreversible inhibitory interaction with ChEs [5, 181]. OPs are substrate analogues to ChEs, which display impressive dissociation constants with their catalytic subunit. The labeled OP diisopropyl fluorophosphate (DFP) was shown to bind covalently to the serine residue at the active esteratic B-site region of ChEs, that is common to all of the carboxylesterases [208, 318]. This property has been used in research aimed at protein-sequencing studies, as well as for testing the developmental effects of ChE inhibition [38, 182] or for behavioral studies of OP poisoning [44]. It should be noted that the binding and inactivation capacity of OPs on ChEs is considerably higher than their effect on other serine hydrolases. Furthermore, even within species the inhibition of specific ChEs by different OPs tends to be highly selective to particular ChE types [21, 169].

In addition to their blocking effect on neuromuscular junctions, OPs also act on the central nervous system, sometimes with devastating effects. It is not yet clear why OPs induce seizure-related brain damage [250] although theoretical models were proposed [268]. In order to improve the designing of therapeutic and/or prophylactic drugs to the short- and long-term effects of OP intoxication, it is hence desirable to reveal and compare the primary amino acid sequence and three-dimensional structure of all of the members belonging to this enzyme family, as well as to the homologous domains in other serine hydrolases. Elucidation of these sequences and their interactions within the ChE molecule and with other elements can deepen the understanding of the mode of functioning of ChEs and the specific amino acid residues involved in this functioning. This has therefore been one of the main directions of research in several groups over the past decade. One of the outcomes of this study has been the analysis of sequence similarities between human AChE and related proteins, based on molecular cloning, DNA sequencing and computer analyses of the derived sequences. This analysis revealed, quite unexpectedly, that the genomic sequences encoding AChE and BuChE in humans do not display a high sequence homology, in spite of the considerable similarity between the protein sequences encoded by these two genes.

2. Molecular Form Heterogeneity in Cholinesterases

ChEs constitute a family of carboxylesterase type B enzymes, all of which catalyze the hydrolysis of choline ester compounds at high rates. The localization of such enzymes at cholinergic synapses, where acetylcholine (ACh) is released and must be rapidly inactivated, can be expected. The AChE enzyme present and concentrated at the neuromuscular junction [79, 80] exhibits a high affinity towards ACh. In addition to AChE it has been demonstrated that a high activity of butyrylcholine hydrolyzing enzyme was present both in the serum [8] and in cholinergic synapses [101]. This enzyme, named BuChE, is also expressed in many additional cell types, including multiple types of embryonic and tumor cells as well as hepatocytes, muscle fibers, endothelial cells and lymphocytes [270]. The role of this enzyme in all of these sites is completely unknown. In addition to ACh, it is capable of hydrolyzing chemically related substrates such as succicurarium and various local anesthetics based on choline esters and related compounds [86, 348]. Because of its high concentration in the plasma and its broad substrate specificity, BuChE is a potentially appropriate scavenger for ecologically occurring poisons that, unless removed at the level of plasma, might get to the brain and induce considerable damage.

In the seventies, different groups [279, 302] studied in detail the molecular polymorphism of both AChE and BuChE in various species, using a variety of biochemical techniques. Several conclusions emerged from this series of studies:

(a) The various globular molecular forms of AChEs and BuChEs include the monomeric, dimeric and tetrameric oligomers of catalytic subunits and are principally composed of catalytic subunits alone [49]. Globular AChE tetramers are subject to selective loss in Alzheimer's disease [18], but it is not yet known what the mechanisms leading to such loss are.

(b) The group defined as 'asymmetric' molecular forms of AChEs includes one triple helical collagen-like catalytic 'tail' subunit associated with one, two or three catalytic tetramers. The collagen-like 'tail' consists of long (50 nm) fibrillary peptides, rich in proline and hydroxyproline residues [280], and is generally assumed to serve as an attachment of its associated tetramers to solid matrix, for example that of the extracellular basal lamina [50].

Several research groups were further able to demonstrate the physical buildup of ChEs at the electron microscopic level, supporting the hypothesis of attachment through the noncatalytic 'tail' subunits. The above model corresponds primarily to the molecular forms of AChEs observed in the electric fishes, namely *Torpedo marmorata* and the electric eel *Electrophorus electricus* [11]. The electric organ of these fishes exhibits very high concentrations of both the nicotinic ACh receptor and of AChE, which is why it served as a highly appropriate model system for studies directed at these molecules. However, it should be noted that the sedimentation coefficients observed for the enzyme in insects [152], amphibians [246], avians [27], or mammals [150, 229, 357] indicate different molecular weight and/or distinct compositions of catalytic and noncatalytic subunits from those previously reported for the *Torpedo* enzyme.

In addition to AChEs, the above model can also be applied for the molecular forms of BuChE, which have very similar structures and sedimentation coefficients as compared with AChE [215, 337, 342]. A different type of enzyme, described in embryonic chicken cells, contains hybrid tetramers associating two dimers of BuChE with two others composed of AChE subunits [340]. This adds a new dimension to the accepted scheme of molecular forms. It also increases the expected level of complexity of molecular polymorphism to be searched for in this enzyme family.

Apart from the different composition of subunits, the molecular polymorphism generally corresponds to different tissue specifications and subcellular localizations. ChEs are produced in fetal, adult and neoplastic brain tissues [110] including noncholinergic areas such as the cerebellum [109]. The enzymes can be either soluble in the cytoplasm, bound to membrane structures [281], or associated with the extracellular matrix material, all according to their tissue origin. The mechanisms responsible for this complex heterogeneity have not been pursued so far.

It has been demonstrated that the dimeric forms of AChEs contain a short C-terminal domain that may be removed by papain digestion [108] and are generally bound to the plasma membrane [236] or to the endoplasmic reticulum, through a phospholipid link [175] which binds this hydrophobic domain of the protein [129–131, 148] and operates via a specific signal peptide [57]. In addition, the same molecular forms of the enzyme are also present as cytoplasmic soluble proteins. The higher oligomeric forms, that mainly appear as ectoenzymes, may be bound to the plasmatic membrane through a highly hydrophobic 20-kd subunit composed of lipid components. This subunit has been found in the bovine brain [163]. The tetrameric enzyme can also be low salt soluble as in the cerebrospinal fluid, where it is the major AChE form, and in the serum where the soluble BuChE tetramer is predominant. Nothing is known as yet regarding the mode of attachment of BuChE catalytic subunits to solid support.

The asymmetrical forms of AChE are assumed to be associated with the fibrillar molecules of the basal lamina [12] through numerous ionic bonds. This assumption is supported by the observation that these tailed forms become soluble at high ionic strength (e.g., 1 *M* NaCl or 0.5 *M* MgCl) and may be dissociated from their solid support and kept in solution. A small proportion (about 20%) of the asymmetrical form remains associated with lipid membranes even in the presence of salt and may be solubilized by the addition of detergent [131, 303]. Several comprehensive reviews discussing the molecular form heterogeneity in ChEs have appeared [23, 225, 333, 336, 338].

In humans, the ChEs in different tissues also exhibit a high degree of polymorphism, as each expresses a different pattern of molecular forms. Detection of high AChE levels in the amniotic fluid is accepted as a diagnostic in neural tube closure defects [40, 47] or for congenital skin disorders [34]. In the liver, which is the presumed source of plasma BuChE, the monomeric and dimeric forms of BuChE are detectable and predominant [Soreq et al., unpubl. data]. The external surface of the red cell has long been known to be extremely rich in the AChE dimeric form [254], and to a lesser extent in the monomeric form [255]. Psychiatric stress, among other causes, affects the level of erythrocyte AChE [127].

In the human fetal brain, the main ChE form is a membrane-bound amphiphilic tetrameric AChE [133, 134, 357]. This form represents about 90% of the total activity, excluding the serum activity [271]. It is largely bound to the external surface of the neurons and is assumed to be presynaptic [221]. A small amount of 16 S has been detected, but it represents only 1–2% of the activity [276]. BuChE [19], mainly as tetramer, has also been demonstrated [357]. Preferential transcription of AChE mRNA over BuChE mRNA transcripts was found in fetal human cholinergic neurons, but not in noncholinergic areas such as the developing cerebellum [22]. The cerebrospinal fluid is very rich in soluble 10 S tetrameric AChE, which is probably secreted by the neurons [70, 308].

AChE activity in the cerebrospinal fluid is subject to drastic decreases in Down's syndrome [354] and in Alzheimer's disease [13, 77, 288], similar to the levels of ACh synthesis [137], whereas AChE in the hypophysis decreases following dehydration and is released by stimulation of the pituitary stalk [6]. In the spinal cord the amphiphilic 10 S AChE tetramer is predominant, and its concentration changes under fetal stress conditions [Dreyfus et al., unpubl. data].

In the neurons of the peripheral autonomous nervous system [138], all of the molecular forms of both AChE and BuChE are more or less detectable [103, 165]. This is also the case of the muscle fibre, where, in addition, ChEs are mainly concentrated at the neuromuscular junction [102] and at the myotendinous junctions. In cultured nerve cells, cholinergic properties develop concomitantly with AChE activity [24].

3. Microinjected Xenopus Oocytes as a Heterologous Expression System to Study Cholinesterase Biosynthesis

Several key steps in the biosynthetic pathways responsible for directing the production of the multiple ChE forms have not been elucidated to date. These include posttranscriptional and posttranslational processes, glycosylation and particle segregation patterns along neurites [286] or in other membranous sites. In order to initiate an experimental approach to the molecular mechanisms underlying the biogenesis of this heterogeneous family of enzymes, a full-length cDNA clone coding for human BuChE

[264] was subcloned into the SP6 transcription vector [187]. Synthetic polyadenylated BuChE mRNA was transcribed in vitro and microinjected into *Xenopus* oocytes, where the translation of tissue-extracted ChE mRNA's has previously been demonstrated [311, 312].

Xenopus laevis oocytes have proven to be a valuable in vivo expression system for the production of a variety of biologically active membrane proteins from synthetic and tissue-derived mRNAs [313]. Proteins extensively studied in the oocyte system include the ACh receptor [26, 234], peptide and amino acid neurotransmitter receptors [160], and various channel proteins [85]. In advanced studies, *Xenopus* oocytes have been used in conjunction with site-directed mutagenesis to investigate structure-function relationships in specific polypeptides.

In our hands, the oocytes produced active BuChE displaying enzymatic properties characteristic of the native enzyme [105, 319]. Coinjection of the synthetic BuChE mRNA with total poly(A)$^+$ RNA from brain and muscle was then employed to examine the involvement of additional tissue-specific factors in the assembly and compartmentalization of the enzyme in the oocytes.

4. The Use of Genetic Engineering to Re-Examine the Immunochemical Cross-Reactivity of Anticholinesterase Antibodies

ChEs are rather large proteins, composed of catalytic subunits of ca. 70 kd each. Clearly, the way such polypeptides are folded is bound to play a pivotal role in their biochemical properties. Previous attempts to reveal the molecular origin for the structural heterogeneity of ChEs were based on the elicitation of polyclonal and monoclonal antibodies against minute quantities of highly purified AChE, prepared from the electric organ of *Torpedo* [97, 240, 275], mammalian brain tissue [222, 233, 269] or from red blood cell membranes [45, 119, 309]. The antibodies produced interacted with all of the molecular forms of either AChE or BuChE, respectively. However, antibodies elicited against AChE did not cross-react with BuChE and vice versa [45, 192, 222]. This was generally interpreted to indicate sequence dissimilarities between AChE and BuChE [228]. On the other hand, monoclonal antibodies with significant cross-reactivity have been seen by at least one group [97]. Also, cDNA cloning [228, 264] revealed 53% homology between the amino acid sequence of the human serum enzyme and that of *Torpedo* AChE [294] and, later on, with human AChE [381], including several identical regions of at least 10 successive amino acid residues. This strongly suggests a common ancestral origin for the two genes encoding these enzymes, which could indicate that the lack of immunological cross-reactivity between AChE and BuChE is not due to a lack of sufficient homology but reflects structural differences. For example, distinct folding patterns of the polypeptide chains could mask homologous regions, or particular glycosylation chains could have the same effect. Another possibility is that the homologous regions are those demonstrating low immunogenicity. To examine these possibilities one would need to elicit antibodies against specific regions of the nascent polypeptide chains that show the greatest homology between various classes of ChE, or which display low immunogenicity.

For this purpose, the N-terminal part of the human BuChE protein, which displays the highest sequence homology to *Torpedo* AChE (amino acid residues 1–198 [see 315]), was produced in bacteria from a recombinant DNA plasmid containing 760 nucleotides from the cloned BuChE cDNA. Antibodies were elicited against this polypeptide and were shown to cross-react with specific molecular forms of both AChE and BuChE from various human tissues.

5. Autoimmune Antibodies to Cholinesterases and Their Clinical Implications

In the neuromuscular junction, various types of conduction defects are known to induce muscle weakness. These include the impaired release of the neurotransmitter ACh, including the Eaton-Lambert syndrome [113], and the disturbed interaction between ACh and the nicotinic ACh receptor, occurring in myasthenia gravis [205]. Severe muscular weakness can also be caused by excessive stimulation, resulting from the accumulation of ACh within the synaptic cleft. This can happen due to inhibition of the neurotransmitter-hydrolyzing enzyme, AChE, as is the case of OP intoxication [181]. Inhibition of AChE in the neuromuscular junction profoundly modifies neuromuscular transmission [305], as has been shown by electrophysiological analyses [193], by studies of the muscle response to nerve stimulation [345], and by observations on spontaneous muscular activity in vivo [118]. In principle, antibodies blocking the activity of AChE in neuromuscular junction should have similar effects.

As mentioned above, both monoclonal and polyclonal antibodies have been raised, for research purposes, against AChE from a long list of species including humans [45, 119, 222]. In many cases, such antibodies presented a strong cross-species and form homology. For example, antibodies raised against both human erythrocyte AChE and *Torpedo* electric organ AChE interact with recombinant human BuChE peptides [104]. However, there are also reports on anti-AChE antibodies which distinguish between two forms of AChE derived from a single tissue [97]. This is because different forms of ChEs share structurally common domains but also contain distinct regions specific to the particular forms. The spontaneous appearance of anti-AChE antibodies with preferential reactivity for muscle membrane AChE in a patient may serve as an in vivo example for the natural occurrence of this phenomenon and its clinical significance. These antibodies appeared in the serum of a patient manifesting severe diffuse muscular weakness. The findings suggest that an autoimmune response to AChE plays a role in the pathophysiology of neuromuscular dysfunction. However, the problem may not be limited to autoimmune antibodies directed specifically against AChE, but includes reactions against homologous proteins.

Several studies [239, 331] have implied that thyroglobulin (Tg), or an immunogenically Tg-like protein, may be an antigen common to both thyroid and eye orbit. Furthermore, there is evidence that in Graves' disease, the protein eliciting an immunological response in the orbit muscle is not Tg itself [178]. Molecular cloning studies revealed a significant homology between the carboxyl terminal of Tg and the N-terminal half of *Torpedo* AChE [294] and human BuChE [264]. Furthermore, a comparison of their hydropathy profiles [315] and the conservation of cysteine residues involved in disulfide bonds [209, 217] suggest that the two proteins may assume a similar tertiary structure and may share common stereoepitopes. It has been proposed that this homology may explain some pathological symptoms observed in Graves' ophthalmopathy [212, 213], including the lymphocytic infiltration seen in the extraocular muscles. To test this hypothesis, the possibility of cross-reactivity between antibodies to Tg and ChE was investigated. For this purpose, IgGs from patients suffering from Graves' ophthalmopathy were interacted with polyclonal antibodies to both proteins in dot blots and protein electrophoretic blots in which the N-terminal part of recombinant human BuChE, which displays particularly high homology to Tg, was compared with the conventionally prepared human Tg. Furthermore, in situ studies have been performed to determine whether patients' IgGs would bind to end-plate regions of muscle, which are rich in ChE activity, as demonstrated cytochemically. The results are indicative of a causal relationship in this pathological state as well, further extending the autoimmune responses against ChEs.

6. Expression of Cholinesterase Genes in Human Oocytes Examined by in situ Hybridization

The involvement of cholinergic mechanisms in oocyte growth and maturation [115, 313] and sperm-egg interaction [272] has been a subject of contention. Muscarinic ACh receptors were detected in the cell membranes of oocytes of mouse [115], monkey, rabbit and man [116] as well as in *Xenopus* frogs [84, 189], where ACh-induced signal responses regulate the metabolism of phosphoinositides and mobilization of intracellular calcium ions [253]. ACh, found in mammalian sperm cells, induces polyspermy in sea urchin eggs [272], whereas a cholinergic antagonist (QNB) prevents fertilization in mouse [125]. Furthermore, the ACh-hydrolyzing enzyme, AChE, which terminates ACh responses, was detected in *Xenopus* oocytes [145, 311, 313]. However, biochemical determinations could not reveal whether AChE is synthesized in the oocyte itself or in surrounding cells [145]. This issue is important for human in vitro fertilization [111], which might involve cholinergic processes. To clarify this question, mRNA presence was pursued in the oocytes. Pronounced expression of human BuChEmRNA was observed in developing oocytes by in situ hybridization, supporting the notion that cholinergic signalling in human oocytes may function independently of surrounding follicular cells, and controls processes of intrinsic importance for oocyte growth and development.

7. Could Organophosphorous Poisons Induce Germline Amplification of Cholinesterase Genes in Humans?

Mammalian cell cultures and tumor tissues often achieve resistance to drugs or inhibitors by selection of cells with inheritably amplified genes producing the target protein of such agents [293, 325]. The ubiquitous enzyme BuChE is strongly inhibited by OP compounds [21] such as war agents or, more commonly, the degradation products of agricultural insecticides such as parathion (*p*-nitrophenyl diethylthionophosphate) [305]. Individuals with defective CHE genes [158] are particularly sensitive to such compounds, probably because they lack the serum BuChE activity, which serves as a natural scavenger to remove anti-ChE compounds in individuals with the wild-type protein. In essence, overproduction of BuChE, particularly in individuals expressing the defective phenotypes, could improve the scavenging process and overcome the poisoning effect. During the course of our research we found an individual expressing a defective serum BuChE with a de novo ca. 100-fold amplification [265] within a genomic DNA fragment hybridizing with BuChE cDNA and localized on chromosome No. 3, the original site where we mapped the CHE gene [316, 317]. A similar amplification was found in one of his sons, suggesting that it initially occurred very early in embryogenesis, in spermatogenesis or in oogenesis.

8. Putative Implication of Cholinesterases in Hemocytopoiesis

In addition to its presence in the membrane of mature erythrocytes, AChE is also intensively produced in developing blood cells in vivo [258] and in vitro [54]. AChE activity was observed in leukemic cell lines [3] and serves as an accepted marker for developing mouse megakaryocytes [56]. Furthermore, administration of ACh analogs like carbamylcholine as well as ChE inhibitors like physostigmine has been shown to induce megakaryocytopoiesis and increased platelet counts in the mouse [55], implicating AChE and/or BuChE in the commitment and development of these hematopoietic cells.

Using the cloned cDNAs coding for both human BuChE [264] and AChE [318], we have localized BuChEcDNA-hybridizing sequences to chromosome sites 3q21, 26 and 16q12 [316, 317] and revealed that the CHE gene does not include AChE-encoding sequences [Gnatt et al., unpubl.]. The 3q21-qter region, where the CHE gene resides, has been found to be subject to variable aberrations in vivo. These include a duplication of the 3q21-qter region associated with a p25-qter deletion and a pericentric inversion of the p25q21 region [7]. Moreover, the chromosome 3q21-26 region includes breakpoints that were repeatedly observed in peripheral blood chromosomes from patients with acute myelodysplastic leukemia (AML) [32, 327]. These cases all featured enhanced megakaryocytopoiesis, high platelet count and rapid progress of the disease [262]. A growing flux of recent reports implicates chromosomal breakpoints with molecular changes in the structure of DNA and the induction of malignancies [37]. Therefore, the connection between: (a) abnormal control of megakaryocytopoiesis in AML as well as in mouse bone marrow cells subjected to ChE inhibition; (b) ChE genes location on the long arm of chromosome 3, and (c) chromosomal aberrations in that same region in AML, has appeared to us as more than coincidental (see Lapidot-Lifson et al. [191] for discussion of this issue).

In order to examine the putative correlation between the human genes coding for ChEs and the regulation of hematopoiesis, or more specifically, megakaryocytopoiesis, we initiated a search for structural changes in the human ACHE and CHE genes from peripheral blood DNA in patients with leukemia, platelet count abnormalities, or both. Our findings demonstrated a significant coamplification of both the ACHE and CHE genes in peripheral blood cells in patients with leukemia and/or abnormalities in their platelet counts, and strongly suggest an active role for these enzymes in the progress of human hemocytopoiesis.

II. Molecular Biological Approach to the Study of Cholinesterase Genes

1. Sequence Similarities between Human Cholinesterases and Related Proteins

The cDNA sequence encoding human BuChE was the first mammalian ChE to be cloned [228, 263]. This was performed using synthetic oligodeoxynucleotide probes designed according to the peptide sequences that were earlier deciphered for *Torpedo* AChE [294]. To search for cDNA clones encoding human AChE, oligodeoxynucleotide probes were synthesized [223] according to the amino acid sequences in evolutionarily conserved and divergent peptides from electric fish AChE [294] as compared with human serum BuChE [209, 263, 264]. These synthetic oligodeoxynucleotide probes, based on codon usage in humans [65, 194], were used for a comparative screening of cDNA libraries from several human tissue origins, since they do not support the formation of mismatched hybrids [51].

Previous biochemical analyses revealed that in the fetal human brain, the ratio AChE:BuChE is close to 20:1 [357]. In constrast, we found the cDNA library from fetal human liver to be relatively rich in BuChE cDNA clones [264]. We therefore searched for cDNA clones that would interact with selective oligodeoxynucleotide probes, designed according to AChE-specific peptide sequences, in various cDNA libraries from fetal brain origin. In particular, we searched libraries from brain basal ganglia that are highly enriched with cholinoceptive cell bodies. Positive clones were then examined for their relative abundance in brain-originated cDNA libraries, as compared with liver. Brain-enriched cDNAs were further tested for their capacity to hybridize with the OPSYN oligodeoxynucleotide probes, previously designed according to the consensus amino acid sequence at the active esteratic site of ChEs [315]. Characterized clones were fully sequenced and subjected to analysis of their detailed properties.

DNA sequence analysis followed by computerized alignment of the encoded primary amino acid sequences of human AChE and BuChE demonstrated, as expected, that the functional similarity among ChEs reflects genetic relatedness. The active site peptide of human AChE, as

Hu. BuChE	NPKSV - -TLFGE*SAGAASVSLH
Hu. AChE	DPTSV - -TLFGESAGAASVGMHL
Tor. AChE	DPKTV - -TIFGESAGGASVGMHI
Dros. AChE	NPEWM- -TLFGESAGSSVNAQL
Dros. Est. 6	EPENV- -LLVGHSAGGASVHLEM
Pig elastase	-GNGVRSGCQGDSGGPLV- -CQK
Bov. chymotrypsin	-ASGV-SSCQGDSGGPLV- -CQK
Bov. prothrombin	EGK-RGDACEGDSGGPFVMKSPY
Bov. factor X	DTQPE-DACQGDSGGPHV- -TRF
Hu. plasminogen	G- -T- -DSCQGDSGGPLV- -CFE
α-Lytic protease	IQTNV-CAEPGDSGGSL
Active site homology of HuAChE to:	Overall homology of HuAChE to:
Hu. BuChE 85%	Hu. BuChE 51%
Tor. AChE 78%	Tor. AChE 56%
Dros. AChE 52%	Dros. AChE 31%

Fig. 1. Active site homologies. Comparison of ChE-active site region sequences with other serine hydrolases. The star indicates ^{3}H-DFP-labeled or active site serine. Amino acid sequence data were based on DNA sequencing of human AChE cDNA clones [Soreq et al., in preparation] and follow Schumacher et al. [294], Prody et al. [263], Hall and Spierer [153] and Myers et al [241] regarding the other active site sequences. Note the considerable difference between the levels of sequence similarities within the ChE family (upper part) and other serine hydrolases (lower part) [reproduced from 318].

deduced from the AChE cDNA clones, revealed 17 out of 21 amino acid residues identical to those of either human BuChE or *Torpedo* AChE (fig. 1). A lower level of similarity (12 out of 21 amino acid residues) was observed in comparison with *Drosophila* AChE [153]. Esterase 6 from *Drosophila* [241], which catalyzes the synthesis of a sex pheromone in females [220], displayed 10 identical residues out of these 21, and several serine proteases – 3 or 4 identical residues only (fig. 1). This comparison draws a distinct line between serine proteases and the family of carboxylesterases, and more particularly, the highly conserved ChEs. Interestingly, there is no homology between CHEs and the ACh receptors at the level of primary amino acids [247], in spite of their common sites for binding ACh.

Once a cDNA sequence is available, it may readily be translated into its inferred amino acid sequence. This method is relatively simple as compared with the determination of the same protein sequence by peptide sequencing. It requires very small quantities of easily amplifiable DNA, while protein sequencing is dependent on the availability of considerable amounts of the highly purified protein. Furthermore, DNA sequencing is highly accurate and subject to fewer errors than peptide sequencing. Indeed, the amino acid sequences of *Torpedo* AChE [294], *Drosophila* AChE [153] and human AChE [318] were all deciphered from their corresponding cDNA sequences.

The coding region in human AChE cDNA and the inferred amino acid sequence of the human AChE protein were compared by computerized analysis [266] with the parallel sequences of human BuChE cDNA [228, 263, 264], and also with the amino acid sequence of AChE cDNA from *Torpedo* [294] and of the more evolutionarily remote AChE cDNA from *Drosophila* [153]. This analysis revealed several peptide regions and DNA sequence domains that are highly conserved in all of the ChEs, particularly at the N-terminal part of the proteins. It further displayed clearly the higher level of divergence between human and *Drosophila* AChE as compared with the extensive similarities between human AChE and BuChE and *Torpedo* AChE. A higher level of conservation was found for all of these proteins and cDNAs at the amino acid level than at the nucleotide level [99]. This was in complete agreement with previous observations on this gene family and its protein products [264, 315]. Significant homology was also observed with the DNA and the amino acid sequence of bovine Tg, in corroboration of previous findings [294, 315].

To further examine the molecular properties of the human AChE protein encoded by the newly isolated cDNA clones, we subjected it to hydrophobicity analysis [159]. The results of this analysis were compared with parallel analyses of the homologous sequences of human BuChE, *Torpedo* AChE and *Drosophila* AChE as well as bovine Tg [212, 213]. Hydrophobicity patterns predicted by this analysis reveal, in all four cases, putatively globular proteins with very short regions of limited hydrophobicity that appear in the same highly conserved positions in the entire family.

In order to search for specific conserved domains in human AChE that could be potentially involved in its hydrolytic activity, the above data were combined with generally accepted concepts on the catalytic functioning of carboxylesterases and of serine proteases. Several serine proteases have been subjected to site-directed mutagenesis [e.g. see 81]. In corroboration of previous enzymology studies, these experiments have demonstrated beyond doubt that three key residues are involved in the charge relay mechanism of serine proteases, donating protons to a peptide bond which is subsequently hydrolyzed. These include the active site serine, a basic histidine residue and an acidic aspartate residue.

In most of the serine proteases, the three key residues appear in the order of His-Asp-Ser with an average distance of 43 and 91 residues between the His and the Asp and between the Asp and the Ser, respectively [88, 89]. Each of these key residues is embedded in highly conserved peptides, 8–18 amino acids in length. The sequence similarities in the peptides surrounding the reactive site serine 198 are cited in figure 1. Careful analysis revealed an invariant aspartate at position 170, that is also surrounded by a highly conserved domain, as expected from residues playing important roles in hydrolytic activity. In spite of the nonconserved distance between this Asp 170 and Ser 198, these two residues appear to be very good candidates for the putative key functions in the charge relay system.

The high pH dependence of the catalytic activity of ChEs [260] and their sensitivity to chemical agents that modify imidazole groups [283] suggest that a histidine residue is also involved in the charge relay mechanism of ChEs. However, there is no conserved histidine on the amino-terminal site of the reactive serine. On the other hand, a highly conserved peptide including an arginine residue can be found around position 147. Arginine replaces histidine in the charge-relay system of phosphodiesterase ST [278], suggesting this residue as a substitute for the conservative His.

An alternative possibility suggests that histidine residues in other positions take part in the charge-relay system of ChEs. Indeed, highly conserved peptides that include histidine residues may be found at positions 423 and 438. Both were suggested [300, 318] to take part in the charge-relay mechanism. According to this proposal, the basic histidine residue would therefore be located on the carboxy side of the reactive serine. An example for reversed positions of the Asp and His residues relative to the Ser may be found in another serine protease, subtilisin [60].

Site-directed mutagenesis studies will be required to distinguish between the above discussed possibilities for putative residues in human AChE. However, in all cases, ChEs are indicated to have a different charge-relay system from that of serine proteases, differing in the identity of the basic residue, in its distance from the reactive Ser or in its relative position on the primary sequence. The possible combinations for the charge-relay key residues of ChEs, including the peptide similarities, are presented in figure 2.

2. Isolation and Characterization of BuChE cDNA from Fetal and Adult Tissues

The results of our cloning studies clearly demonstrated that within the fetal brain and liver, AChE and BuChE are produced from two distinct mRNA transcripts. However, it could not answer the question whether

A–C. Key residues combinations

A. Arg 147 – Asp 170 – Ser 198 (23 – 28)
B. His 423 – Asp 170 – Ser 198 (253 – 28)
C. His 438 – Asp 170 – Ser 198 (268 – 28)

D. Sequence similarities

	Arg 147 region	Asp 170 region	His 423 region	His 438 region
Hu. AChE	yRvgafgflal	nvgllDqrlal	Hrastlswmgv	pHgyeieftfgftfg
Hu. BuChE	yRvgalgflal	nmglfDqqlal	Hrssklpwmgv	mHgyeiefvfgfvfg
Tor. AChE	yRvgafgflal	nvgllDqrmal	Hrasnlvwmgv	iHgyeiefvfgfvfg
Dros. AChE	yRvgafgflhl	nvglwDqalai	Hrtstslwmgv	lHgdeieffg yffg

Fig. 2. Putative residue combinations for the charge-relay system of the ChE catalytic sites. *A–C* Possible combinations for the key residues in the charge-relay system of human ChEs are presented. Residues are numbered according to their appearance in the mature human BuChE protein. The number of residues between the first and second pairs of amino acids are marked in parentheses for each combination. *D* Sequences were aligned as previously detailed [264, 315] for human AChE (Hu.AChE), human BuChE (Hu.BuChE), *Torpedo* AChE (Tor. AChE) and *Drosophila* AChE (Dros. AChE). The position of each putative catalytic site residue within the surrounding 11 amino acid sequence is shown above each region. Residue numbering begins with 1 as the first amino acid of the mature human BuChE protein [209], since the human AChE sequence is deduced from cDNA data only. The putative catalytic residues are shown in capital letters [reproduced from 318].

either AChE or BuChE are produced from different transcripts in fetal as compared with adult tissues. Alternative splicing has so far been demonstrated only for *Torpedo* AChE [135, 136, 300, 301]. To approach this issue in humans, cDNA screening was performed in parallel in cDNA libraries from fetal and adult liver. The first probes employed were those coding for BuChE. Thus, we have screened a cDNA library from adult liver origin constructed in a λ gt10 vector in order to isolate cDNA clones coding for BuChE. Two main probes were used to detect the presence of the right cDNA insert by hybridization: (a) The full-length BuChE cDNA clone previously isolated from human fetal brain and liver libraries [264] and which was found to code for the catalytic subunit of BuChE as it appears in the adult serum [209]; (b) A cDNA clone coding for the 200 first N-terminal amino acid of this protein [104, 263].

The assumption was that the coding region could not be subject to large variations, since the protein inferred from it remains the same in embryonic and adult tissues. However, we searched for 3′-alternated cDNAs coding for BuChE, in analogy to those for AChE in *Torpedo* [300].

The 5′-partial probe should in principle pick these clones with no difficulty. The results of the screening experiment demonstrated that BuChE cDNA inserts were much less abundant in the adult liver library that in the fetal liver library. This is in apparent contradiction with the observation that BuChE levels in serum increase with postnatal development [348]. Since plasma BuChE is most probably produced in the liver, the high abundance of BuChE cDNA clones in the fetal liver library suggests BuChE synthesis which is unrelated to liver functioning. At the examined gestational age (17 weeks), hemopoiesis is still going on in liver islands. Our current assumption is, therefore, that the CHE gene is transcriptionally active while fetal hemocytopoiesis continues. Only 3 of the putative BuChE cDNA clones out of 5×10^5 were positive after the third screen with labeled BuChE cDNA probe, as compared with 40 in a parallel screen of the fetal liver library. One of the adult liver inserts has been purified and then subcloned into the sequencing single-stranded phage M13.

Nucleotide sequencing of the full-length adult liver BuChE cDNA has been performed using the dideoxy Sanger method [291] as previously described [264]. The nucleotide sequence of the adult liver BuChE cDNA was identical to the fetal sequence in all respects, clearly demonstrating that it was transcribed from the same gene by a similar mechanism of posttranscriptional processing. Figure 3 presents the full-length BuChE cDNA.

In view of the low abundance of BuChE cDNA clones in the adult liver library, our findings do not exclude the possibility that an alternatively spliced species of BuChE mRNA exists; further experiments should be performed to clarify this issue. However, it should be pointed out that the BuChE cDNA sequences derived from the fetal brain and the fetal liver libraries were also identical [264], and indistinguishable from the adult liver sequence described presently. This indicates that the same enzyme is expressed by the fetal and the adult liver and by the brain tissue. This finding, together with the single BuChE mRNA species lighted up in RNA blot hybridization [264], indicates that alternative splicing from the CHE gene probably does not occur in normal fetal and/or adult brain and liver.

The complete amino acid sequence of BuChE reveals some characteristics which are of interest for understanding the biological properties of this protein. The first 70 amino acids represent a fully open reading frame (ORF) containing a leader peptide which may be cleaved off in the course of biosynthesis and allows a transmembranous transfer of the nascent protein during its translation and posttranslational processing. Similarly, ORFs were found in the 5′-domains from *Torpedo* AChE [294] and *Drosophila* AChE [153], suggesting a biological significance for this putative peptide. Therefore, the possibility had been considered that the 5-ORF peptide remains in the complete, mature BuChE molecule, where it serves

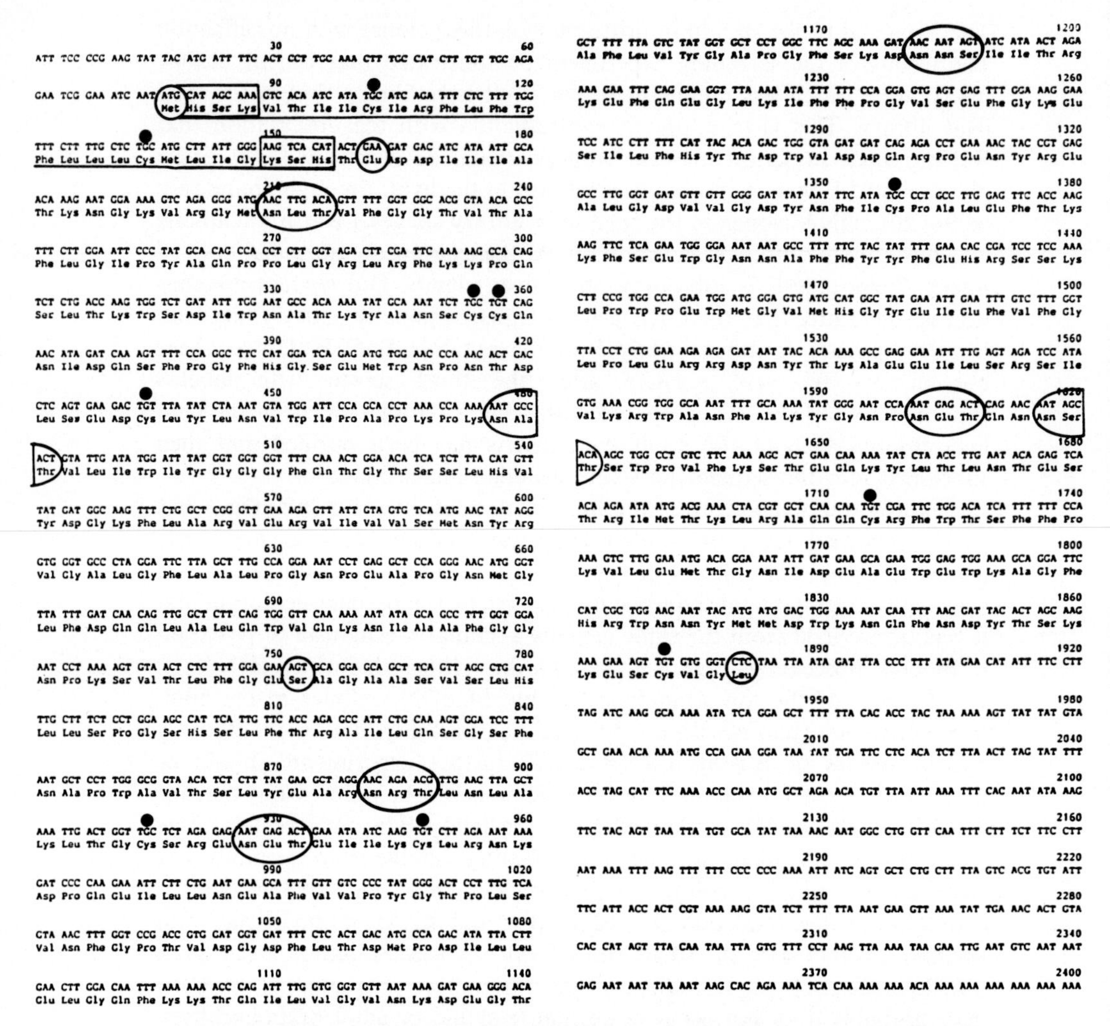

Fig. 3. Primary structure of human BuChE encoded by cDNA from adult liver. The 2.4-kb nucleotide sequence was translated into its encoded amino acid sequence. Nucleotides are numbered in the 5′ to 3′ direction, and the predicted amino acids are shown below the corresponding nucleotide sequence. Underlined are a putative ribosome binding site (nucleotides 30–36) and signal peptide (nucleotides 88–147) [155], with boxed three polar amino acid residues at both ends. Seven potential sites for N-linked glycosylation (starting at nucleotides 208, 475, 880, 925, 1180, 1600, and 1615), predicted by the sequence Asn-Xaa-Thr/Ser, in which Xaa represents any amino acid except proline [30], are ovally encircled. The initiator methionine, active site serine, N′-terminal glutamate and C′-terminal leucine residues are circled. Cys residues are marked by full circles on top. The BuChE cDNA sequence also includes a long 3′-untranslated region, ending with a polyadenylation site and a poly(A) tail. The sequence is identical in all respects to that of fetal BuChE cDNA [264].

as a cytoplasmic domain extending a membrane anchor similar to that of the asialoglycoprotein receptor [322]. Expression of the recombinant BuChE translated from this cDNA in microinjected *Xenopus* oocytes later revealed that it is faithfully cleaved downstream from the signal peptide and transported efficiently. This study ruled out the possibility of functionally active 5′-ORF peptide in BuChE [319]. Further consensus sequences within the BuChE mRNA transcript reflect the occurrence of glycosylations, taking place within the Golgi apparatus. The sequence also shows seven cysteine residues, among them six that are probably implicated in intramolecular S-S bonds. The remaining cysteine is involved in the intersubunit S-S bond that covalently binds BuChE catalytic subunits to each other [209]. Interestingly, the human AChE sequence is considerably richer in cysteine residues but poorer in N′-linked glycosylation sites, which might indicate a different intramolecular folding of its polypeptide backbone and explain the immunochemical differences between AChE and BuChE. The serine residue in the active esteratic site, a major characteristic of the enzyme, has been conserved in all of the members of the ChE gene family [328].

3. In ovo Translation of Synthetic Butyrylcholinesterase mRNA in Microinjected Xenopus *Oocytes*

Several considerations lead us to the next step, namely injection of the mRNA transcribed from BuChE cDNA into *Xenopus* oocytes. Early attempts to express the human BuChE polypeptide in a bacterial expression system [in collaboration with J. Hartman and M. Gorecki, Biotechnology General, Nes-Ziona] resulted in the production of a totally inactive, although immunochemically positive protein [unpubl. data]. Also, in vitro translation of synthetic BuChE mRNA in reticulocyte lysate [259] failed to produce the catalytically active enzyme [297]. We attributed both results to the known effect that only a secretory eukaryotic cell could be expected to be capable of performing correctly the posttranslational S-S bonding as well as glycosylations, both of which never occur in intracellular bacteria-produced proteins. This implied that some properties of the enzyme most probably resulted from its natural mode of processing. Second, the possibility existed that the distinct properties characteristic of AChE and BuChE in terms of their substrate specificity and sensitivity to selective inhibitors were the result of posttranslational modifications, and should not be attributed to their different amino acid sequences. Third, we were interested to examine, by expressing synthetic BuChE mRNA in microinjected oocytes, what are the factors which lead to the numerous molecular forms observed for ChEs in different tissues.

a. Oocytes Produce Active Recombinant Butyrylcholinesterase

The synthetic mRNA we injected was transcribed from a cDNA construct containing the binding site for the SP6 salmonella phage RNA polymerase [187] followed by the cDNA sequence [264] encoding an amino acid sequence which is identical to that of human serum BuChE [209]. In vivo this enzyme may be distinguished from AChE by its substrate preference and sensitivity to selective inhibitors. In our first experiments we therefore assessed the possibility that the primary amino acid sequence is sufficient to confer ligand binding specificity.

The enzyme produced by the oocyte in response to injection of SP6 BuChE mRNA demonstrated a clear preference for butyrylthiocholine (BuSCh) over acetylthiocholine in spectrophotometric activity assays. Enzymatic activity induced by as little as 2 ng of the recombinant mRNA averaged 8.2 ± 1.6 nmol BuSCh hydrolyzed/h – 2 orders of magnitude higher than the activity achieved with total poly(A)$^+$RNA from embryonic brain [312]. By comparison, acetylthiocholine hydrolysis in these experiments ranged 3–4 times longer, averaging 2.2 ± 0.3 nmol/h/ng mRNA. An apparent K_m of 2×10^{-3} *M* BuSCh was determined for the secreted, cytoplasmic, and membrane-bound fractions of the enzyme. This value correlates to values reproducibly calculated for several control human serums in our laboratory. The total BuChE activity measured in injected oocytes was distributed as follows: $4 \pm 1\%$ in the medium, $38 \pm 7\%$ low salt soluble and $58 \pm 7\%$ extracted. The oocyte enzyme exhibited characteristic inhibition by the BuChE-specific OP inhibitor tetraisopropylpyrophosphoramide (iso-OMPA) while resisting inhibition by the AChE-specific quaternary inhibitor 1,5-bis(4-allyldimethylammonium-phenyl)pentan-3-one dibromide (BW284C51) [319].

The IC_{50} values of $2–3 \times 10^{-6}$ *M* iso-OMPA calculated for the secreted and cytoplasmic pools of the enzyme were essentially identical to the value determined in parallel for the human serum enzyme. The IC_{50} value 1×10^{-5} *M* measured for the detergent-extracted enzyme indicates a limited but significant decrease in BuChE affinity for this inhibitor, but the decrease was apparently due to the presence of detergent. All three fractions displayed a 10–20% inhibition by high concentration (10^{-4} *M*) BW284C51.

b. Synthesized Butyrylcholinesterase Dimers Assembled into Complex Multimeric Forms following Coinjection with Tissue RNAs

Linear sucrose gradient ultracentrifugation was employed to examine the nature of subunit assembly in recombinant BuChE. When injected with synthetic BuChE mRNA alone, all three subcellular oocyte fractions contained distinct BuChE activity displaying a sedimentation coefficient of

5–7S and corresponding to the globular dimeric enzyme form. This implied the assembly of nascent BuChE polypeptides, most probably in the oocyte's Golgi apparatus [284, 285]. In contrast, native serum BuChE exists primarily as globular tetramers [208]. Supplemental muscle poly(A)$^+$ RNA induced a complete array of BuChE molecular forms in the membrane-associated fraction of the oocytes, including a heavy 16S peak characteristic of the neuromuscular 'tailed' form of the enzyme [150] and which represents a significant fraction of human fetal muscle BuChE [105]. Coinjection of fetal brain poly(A)$^+$ mRNA with the synthetic BuChE mRNA induced a considerable peak of membrane-associated BuChE activity sedimenting as 12S globular tetramers [319], the primary BuChE form found in human fetal brain [357].

c. Recombinant Butyrylcholinesterase Associates with the Oocyte Surface

In human brain and muscle, BuChE associates with the extracellular surface from which it can be detached by salt and detergent [336, 357]. To examine the mode of association of the recombinant enzyme with the external surface of the injected oocytes, an immunohistochemical approach was taken. Frozen sections of BuChE mRNA-injected oocytes were incubated with antibodies to *Torpedo* AChE, known to cross-react with human BuChE [104] and then with fluorescein-conjugated second antibody. Characteristic green signals on the oocyte surface could be easily distinguished from the yellow autofluorescence emitted from the internal oocyte yolk vesicles [251]. Recombinant BuChE accumulations appeared on the surface of the oocytes in the form of either small round 'clusters', about 5 μm in diameter, or elongated 'patches' of 20 μm in diameter, similar to those observed along avian neurites [286]. Both types of structures were relatively concentrated at the animal as opposed to the vegetal pole [105] and could be identified within 30 min after microinjection. Signal intensity increased with time up to 2.5 h postinjection, at which point maximum intensities were achieved. In contrast, muscle poly(A)$^+$ RNA which contains about 0.001% of BuChE mRNA [H. Soreq, unpubl. observations] created very weak surface-associated signals. Blockage of glycosylation by tunicamycin induced the accumulation of fluorescent signals around intracellular vesicles.

d. Coinjection with Tissue mRNAs Intensifies Surface-Associated Butyrylcholinesterase Signals

Coinjection of brain and muscle mRNA recombinant BuChE mRNA resulted in 2.4- and 3.6-fold increases, respectively, in the total surface area occupied by patches and clusters while maintaining the 2- to 3-fold disproportionate distribution at the animal over vegetal pole. In both cases, the relative intensity of fluorescent staining was also increased although the enhancement obtained in oocytes coinjected with muscle mRNA was significantly more dramatic. Particularly high accumulations of immunofluorescing structures were detected in oocyte sections which included the injection site.

Electron microscopic analysis using second antibodies coupled to 5-nm gold beads confirmed the qualitative differences between the surface-associated BuChE accumulations in oocytes coinjected with brain or muscle mRNA. The high sensitivity of analysis at this level further revealed that these BuChE deposits were not primarily associated with the plasma membrane of the oocyte itself, but in fact were mostly linked closely with the external layer of the extracellular material surrounding the oocytes and their follicle cells. Some gold particles could also be detected at the level of the oocyte microvilli and follicle cells, perhaps caught en route to the cell surface. In tunicamycin-injected cells, immunogold beads were concentrated around intracellular vesicles.

4. Cross-Homologies and Structural Differences between Cholinesterases Revealed by Antibodies against Recombinant Butyrylcholinesterase Polypeptides

In order to search for similarities and differences between various ChEs at the level of the naked polypeptide, the amino acid sequence of human BuChE, as deduced from cDNA sequence information [264], was subjected to computerized analysis by the prediction of Chou and Fasman [67]. Measures of α-helix and β sheet values [68, 91], 'best guess' predictions of immunogenicity [132] and hydropathic characteristics [190] were combined to examine the expected immunogenicity of specific regions within the polypeptide sequence. Using this analysis, the N-terminal 200 amino acids of BuChE were found to be particularly low in immunogenicity. Since this is also the part of BuChE which shows greatest homologies to other ChEs [315], we selected this polypeptide as an antigen. To prepare this protein in a naked form, the FBChE12 insert [263, 264] was used. This insert consists of a fully open reading frame and codes for the signal peptide and the N-terminal 200 amino acids of human BuChE [315]. It was subcloned into the pEX bacterial expression vector [324] and ligated to the 3′-end of the gene encoding β-galactosidase in the pEX plasmid.

a. Partial Recombinant Butyrylcholinesterase Polypeptides Are Immunoreactive with Various Cholinesterase Antibodies

In the absence of insert, shift of bacteria transformed with plasmids to 42 °C induces synthesis of β-galactosidase [324]. In contrast, induction of β-galactosidase expression in bacteria transformed with the pEX_3 FBChE12 construct resulted in the production of a fusion protein of ca. 125 kd in length, which was mostly proteolyzed in the bacteria to yield a series of proteolytic products in the range of 35–75 kd. (It should be noted that fusion proteins produced from cDNA inserts of ca. 400 nucleotides or more in pEX vectors tend to be proteolyzed in the bacteria [324].) Also, the synthesis of the cro-β-galactosidase (110 kd) ceased in these bacteria, as compared with its synthesis in bacteria transfected with the original pEX plasmid [104].

The nature of polypeptides translated from pEX_3 FBChE12 plasmids was examined by immunoblot analysis of bacterial extracts. It was found that all of the proteolytic products derived from the fusion protein reacted with antibodies against bacterial β-galactosidase. Of these, two peptides of 40 and 35 kd in length, but not the heavier ones, interacted specifically in protein blots with rabbit antibodies against AChE from both human erythrocytes and *Torpedo* electric organ. In addition, the 40- and 35-kd polypeptides reacted specifically in immunoblots with the AE1–5 mouse monoclonal antibodies raised against human erythrocyte AChE [119]. Altogether, the interaction with antibodies indicated that these polypeptides are derived from a partially proteolyzed fusion protein of β-galactosidase-FBChE12, with the 40- and 35-kd polypeptides containing the information encoded by the FBChE12 cDNA insert as well as some parts from the C-terminal domain of β-galactosidase. Furthermore, the immunoblot analysis revealed that these BuChE-derived polypeptides share immunological properties with both human and *Torpedo* AChE. Protein blot analysis with antiserum against whole human serum proteins, which efficiently interacts with BuChE in crossed immunoelectrophoresis plates [314], failed to reveal a significant specific interaction, perhaps because the antibodies interacting with BuChE in this complex antiserum only recognize the mature protein.

b. Antibodies Elicited against Recombinant Butyrylcholinesterase Interact with Native Cholinesterases

The 40- and 35-kd protein products synthesized in bacteria from the FBChE12 insert were employed to elicit rabbit anti-ChE antibodies. The resultant rabbit serum, purified by affinity chromatography, was tested by immunoblot analysis against the electrophoretically purified antigen. It was also reacted with blotted purified human erythrocyte AChE, as well as with

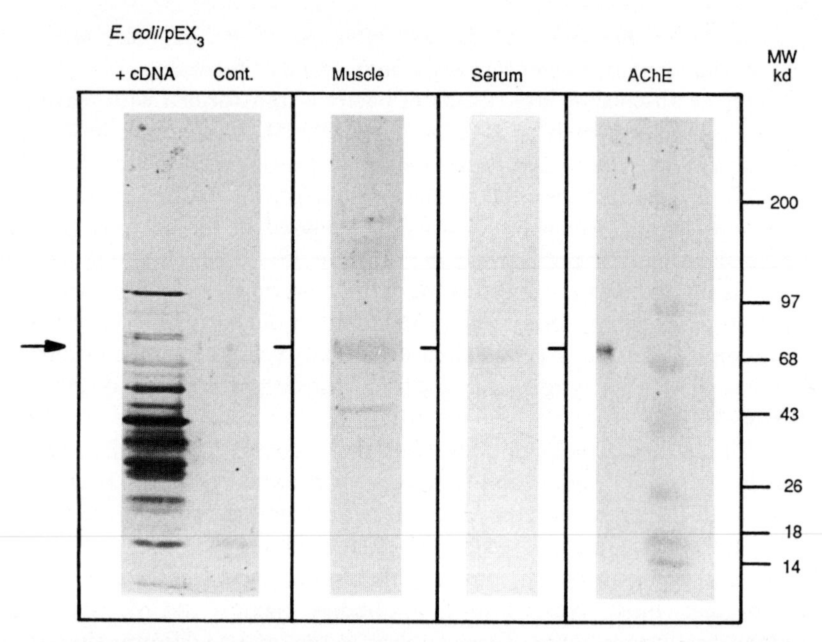

Fig. 4. Antirecombinant ChE antibodies: Immunoblot analysis with antibodies elicited against the cloned fragment of human BuChE. Antiserum against the fragment of human BuChE expressed in pEX_3 FBChE12 plasmids was used for immunoblot analysis in a 1:500 dilution. Proteins were extracted from the pellet of bacteria carrying pEX_3 FBChE12 recombinants (+) or native pEX_3 plasmids (Cont.). Highly purified erythrocyte AChE (20 ng, gratefully received from Drs. E. Schmell and T. August, Baltimore), serum, and total protein extract from fetal human muscle were loaded in parallel [reproduced from 104, with permission].

protein extracts from *Escherichia coli* bacteria transformed with nonrecombinant pEX_3 plasmids and from bacteria transformed with pEX_3 FBChE12 plasmids. The antibodies interacted specifically with the 70,000-kd purified AChE, with similarly migrating proteins in serum and muscle, with several additional muscle proteins of variable sizes and with a set of polypeptides from the pEX_3 FBChE12-transformed bacteria, the largest of which was ca. 95,000 kd in length. This analysis, presented in figure 4, demonstrated that the anticloned BuChE antiserum interacts specifically with the recombinant human BuChE and with highly purified erythrocyte AChE, both under complete denaturation and immobilization conditions.

c. Antibodies to Recombinant Butyrylcholinesterase Interact Preferentially with Particular Globular Forms of Cholinesterases

The rabbit antibodies directed against recombinant human BuChE peptides were used for the immunoprecipitation of native and denatured human serum BuChE and AChE-enriched fractions prepared from red blood cell membranes. For this purpose, serial dilutions of the rabbit antiserum were incubated with the enzyme samples, using immunobeads covered by a goat anti-rabbit IgG (Amersham) as a second antibody. Immunoprecipitated pellets of native ChEs were analyzed for ChE activity and the levels of thiocholine released from acetylthiocholine (for AChE) or butyrylthiocholine (for BuChE) were determined. Approximately 15% of the BuChE and 10% of the AChE activities could be precipitated, with a significantly higher level of BuChE precipitation and with the same and optimal dilution of the antiserum (1:80) for both activities [104]. The denatured ChEs were incubated with ^{3}H-DFP and dialyzed prior to the immunoprecipitation reaction, so that the radioactivity recovered in the immunoprecipitated pellets reflected the efficiency of the immunoreaction with denatured ChEs. In this case as well, precipitation curves were obtained for both AChE and BuChE with the same 15–20% level and optimal dilution of the antiserum.

In order to determine which of the various molecular forms of AChE and BuChE from different tissue sources interact with these antibodies, the relevant tissue extracts were subjected to direct immunoreaction followed by sucrose gradient centrifugation and ChE activity measurements. In this experiment, the antibody-reacted serum BuChE tetramers displayed a sedimentation coefficient of ca. 13.5S, as compared with 11.8S for the native enzyme; in contrast, the erythrocyte dimeric enzyme sedimented with exactly the same rate before and after the immunoreaction. Different results were obtained when the complex activities from fetal muscle extracts were similarly immunoreacted. BuChE dimers, but not BuChE tetramers from fetal muscle, changed their sedimentation coefficient when reacted with the antiserum. When AChE activities from the same extracts were measured, it was found that here as well, the dimeric form displayed an apparent shift in sedimentation. Small but significant shifts were also detected in the sedimentation patterns of the 10S and the 12S forms of muscle AChE. Thus, the sedimentation rates of AChE dimers from fetal muscle, but not from erythrocytes, and BuChE tetramers from serum, but not from fetal muscle, were significantly affected by these antibodies (fig. 5).

d. Interaction in situ of Antibodies to Recombinant Butyrylcholinesterase with Motor Endplate-Rich Regions

Junctional ChEs are particularly concentrated in motor endplate-rich regions. To examine whether the antibodies directed against recombinant BuChE peptides interact with exposed protein epitopes in motor endplates, an in situ technique was developed for the immunocytochemical localization of binding sites for these antibodies.

For this purpose, microdissected bundles of fetal muscle fibers were incubated with the antiserum, washed and further incubated with ^{125}I-protein A (Amersham). Immunocytochemical localization of silver grains, indicating the binding of ^{125}I-protein A to the antigen-antibody complex, was then performed by photographic emulsion autoradiography followed by cytochemical staining of ChE activities. Using this technique, it was found that the labeling patterns obtained with radioactive protein A corresponded to the concentrated sites of ChE activity on the muscle fiber. Only one 'patch' of activity was detected along each fiber, which suggests that it represents the single normal monofocal innervation that is characteristic of the human diaphragm muscle at this developmental stage (20 weeks' gestation). Moreover, both the size of these patches (about 10 μm), and their typical morphology ('en plaque', a simple gutter without secondary foldings) were in good agreement with the properties of the expected immature motor endplates. Normal serum, incubated under the same conditions with similar fetal muscle fibers, did not create any labeling

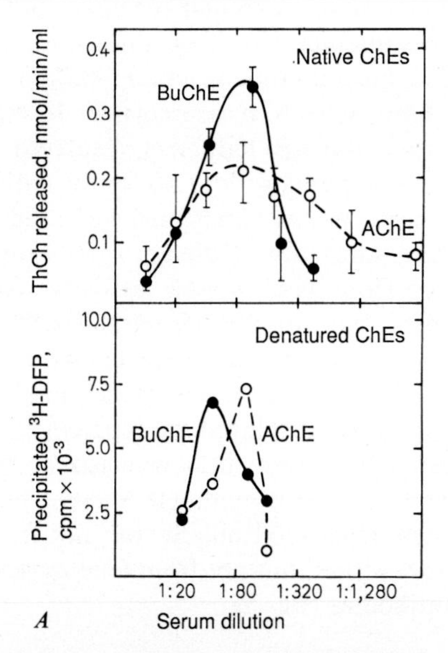

Fig. 5. Interaction of human ChEs with antibodies to recombinant BuChE. *A* Precipitation curves. The rabbit antiserum elicited against recombinant BuChE was serially diluted and incubated with enriched preparations of serum BuChE or erythrocyte AChE, either in their native forms (top, average of three different experiments and deviations) or following interaction with [^{3}H]-DFP and denaturation (bottom). Complexes with the second antibody were formed by further incubation with goat anti-rabbit IgG (BioRad) conjugated to Sepharose beads, and precipitated by centrifugation. The extent of interaction was then determined, either by acetylthiocholine hydrolysis measurements for the native enzyme or by counting the precipitated [^{3}H]-DFP for the denatured ChEs, respectively. Control reactions,

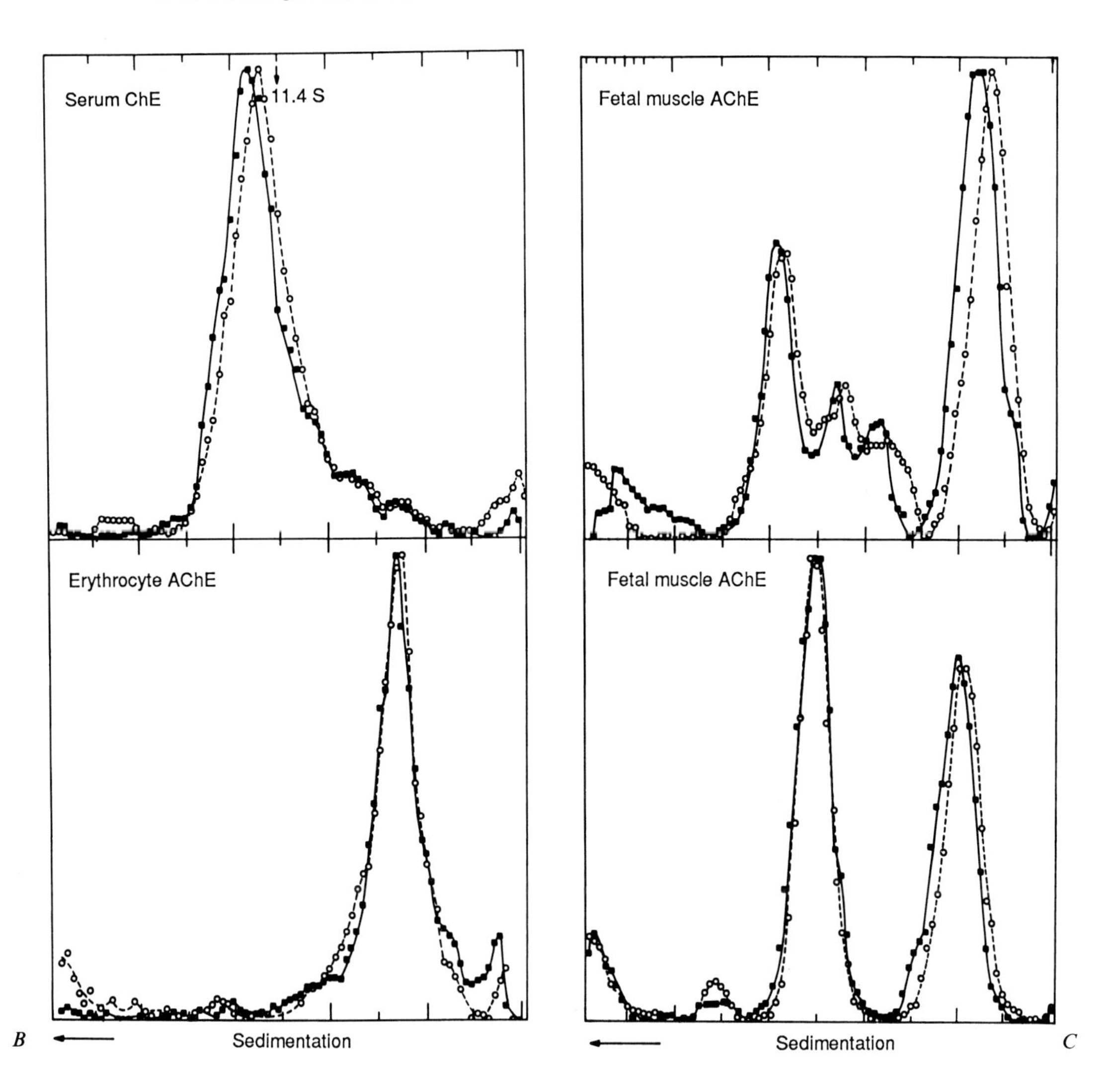

to exclude the possibility of nonspecific binding to the immunobeads, were incubated with nonimmune rabbit serum at similar dilutions. ThCh = thiocholine. *B, C* Sucrose gradient sedimentation. Antirecombinant BuChE antibodies were incubated with enriched serum BuChE or erythrocyte AChE or with fetal muscle extracts, and ChEs were separated from each of these mixtures by sucrose gradient sedimentation (●), in comparison with the native ChE forms (○). Bovine catalase (11.3S) served as the principal sedimentation marker, and ChE activities in separated gradient fractions were determined by hydrolyzing either acetylthiocholine (for AChE) or butyrylthiocholine (for BuChE) by a microtiter plate modification of the Ellman technique [98]. Note shifts in the positioning of serum BuChE tetramers and fetal muscle AChE and BuChE dimers [reproduced from 104, with permission].

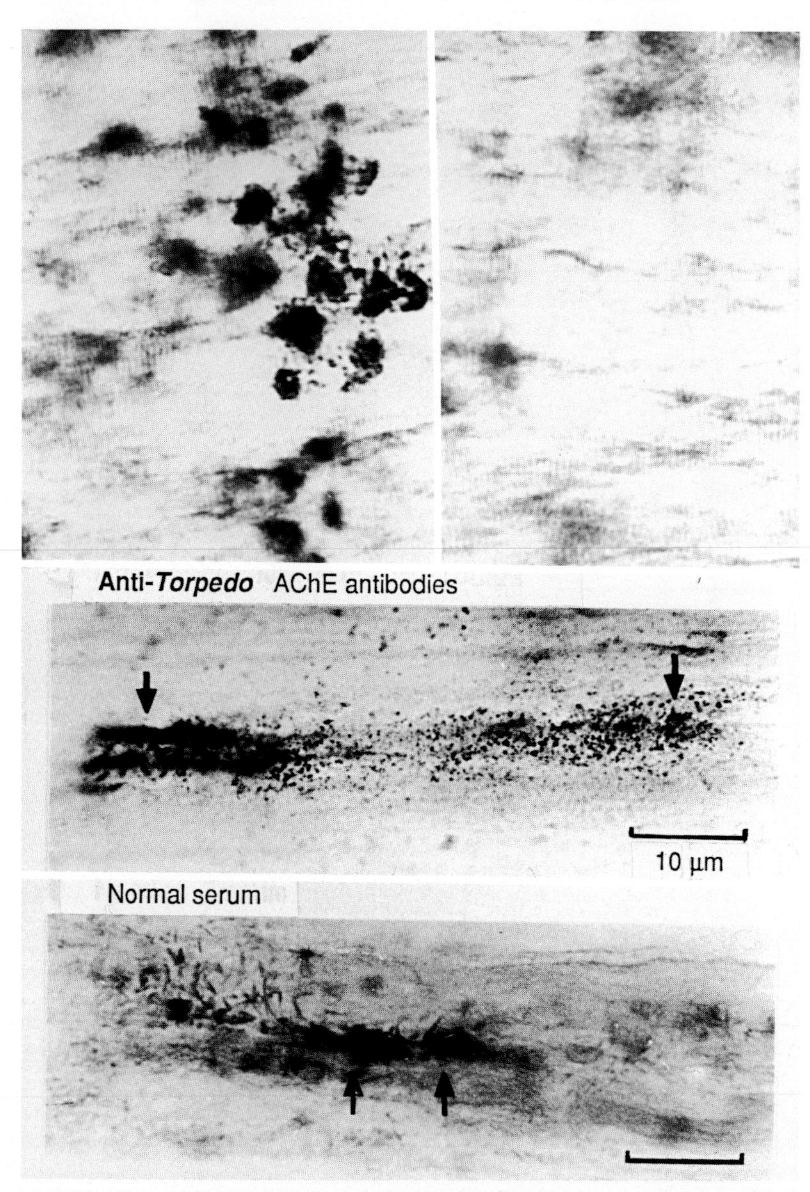

Fig. 6. Binding of antibodies directed against recombinant ChE to crushed fetal muscle fibers. Crushed fibers were incubated with antirecombinant BuChE antiserum (1:100 in PBS) and then with ^{125}I-labeled protein A. These were exposed under Kodak NTB-2 emulsion for 2 days and cytochemical staining was performed according to Koelle and Friedenwald [180] as modified by Karnovsky and Roots [172] and further modified by Koenig and Rieger [183].

at all, while the rabbit antiserum elicited against *Torpedo* electric organ AChE labeled these fibers as well [104] (fig. 6). It therefore appears that the antibodies towards recombinant human BuChE could efficiently interact with solid-phase bound ChEs, either in blots or in fixed tissue, but were much less potent in interacting with soluble ChE activities.

5. Autoimmune Antibodies to Neuromuscular Junction Cholinesterases

AChE has been suggested to play a pivotal role in regulating post-synaptic efficiencies of central nervous system synapses, in primitive organisms such as *Aplysia* [126] as well as in the more complex mammalian nervous system [144]. Chemical inhibition of AChE by OP poisons results in the accumulation of ACh within the synaptic cleft in neuromuscular junctions. This leads to excessive stimulation, neuromuscular dysfunction, and nerve and muscle atrophy [181]. During the past few years we have found low levels of cholinesterases and inhibitory anticholinesterase antibodies in the serum of several patients with symptoms of neuromuscular dysfunction that resembled those of OP intoxication. One of these cases was studied in detail [206].

The patient was admitted because of generalized muscle weakness and dyspnea, abdominal discomfort, diarrhea and weight loss. Blood, urine and cerebrospinal fluid tests were normal except for fluctuating, reduced levels of AChE and BuChE which tended to be particularly low during dyspnea attacks coupled with paralysis [206]. There were no antibodies against the nicotinic ACh receptor; however, precipitating anti-AChE antibodies were found in very high titer. The antibodies were capable of inhibiting and immunoprecipitating AChE from extracts of fetal human brain, muscle, liver, heart, and adrenal, either directly or with the aid of protein A-Sepharose beads. Enzyme inhibition was not affected by protease inhibitors. In contrast, it was abolished following preadsorption of immunoglobulins with goat anti-human Fab, confirming the immunological nature of the inhibitory interaction. Furthermore, the inhibition was particularly strong with membrane-bound AChE (fig. 7). The interacting enzyme was 'true' AChE, as was shown by its resistance to 10^{-5} *M* of the selective

---

Note appearance of silver grains, representing binding of labeled protein A to antibodies, at the area rich in endplates revealed by high ChE activity (top left) but none in endplate-free regions (top right). Anti-*Torpedo* AChE antibodies (gratefully received from Dr. P. Taylor, San Diego) induced parallel binding (middle) whereas normal preimmune serum did not induce any labelling at all (bottom) [reproduced from 104, with permission].

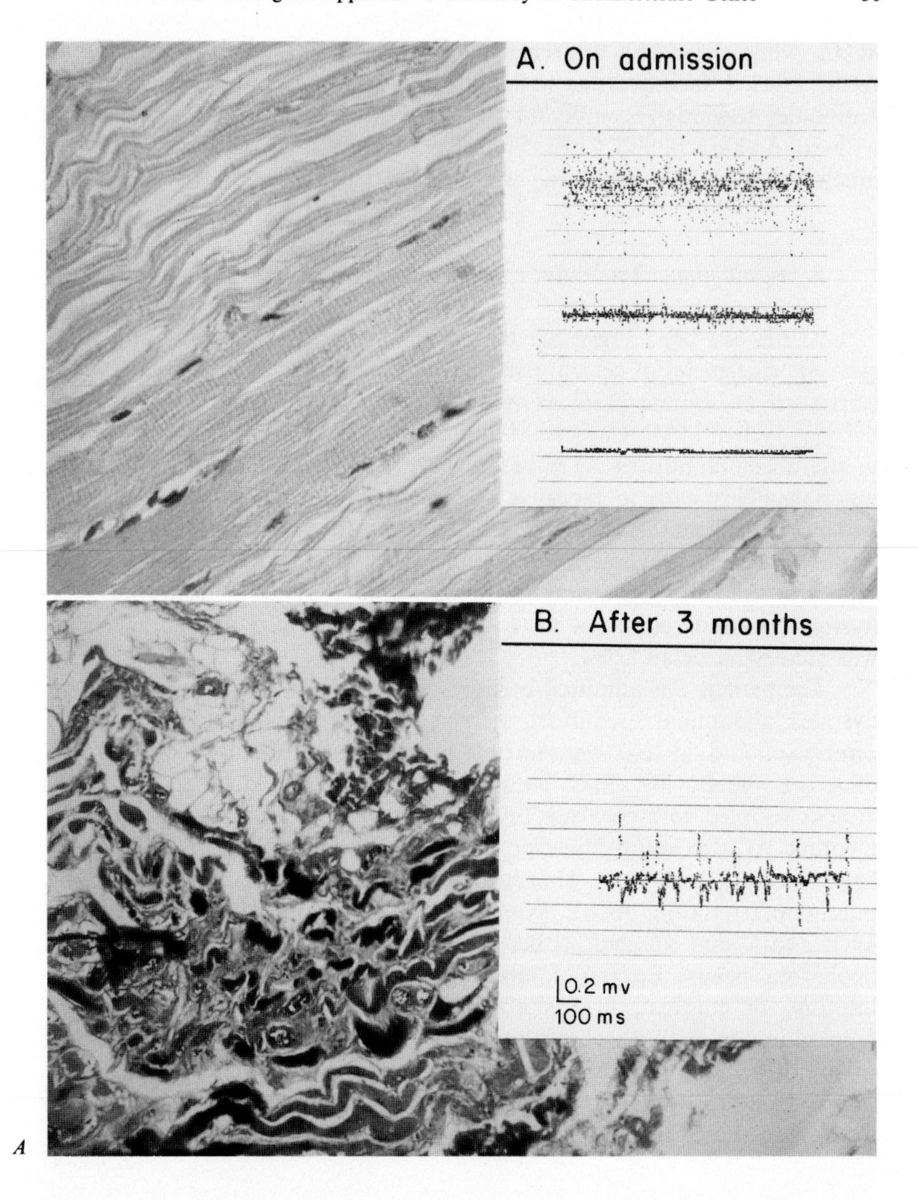

Fig. 7. A Autoimmune anti-AChE antibodies associated with nerve and muscle atrophy and decremental electromyographic response. Biopsies of the deltoid muscle (up) and sural nerve (down) were performed 3 months after admission of patient due to neuromuscular dysfunction. Note the shortened and tortuous muscle fibers, some of which are hyalinized (darkened areas), abundance of connective tissue, and infiltration of fatty tissue. Degenerative changes in the sural nerve are manifested by uneven distribution of the elongated Schwann

inhibitor iso-OMPA, which blocks BuChE activity [274]. Also, radioiodinated purified AChE from human erythrocytes was precipitated efficiently by the examined serum.

To examine the tissue-specific preference of these anti-AChE antibodies, AChE from various fetal tissues was separated into globular, buffer-soluble forms and asymmetric forms, extractable by salt + detergent [357]. Direct immunoprecipitation of the enzyme was carried out with serial dilutions of the examined serum and with normal human serum for control. High preference for the immunoprecipitation of the membrane-associated fraction of muscle AChE was observed [206]. This was associated with pronounced degeneration of muscle and nerve, which coappeared with decremental electromyography records (fig. 7). The differential interaction could occur with a common immunogenic epitope, present in all ChE forms but better exposed in muscle membrane AChE. Alternatively,

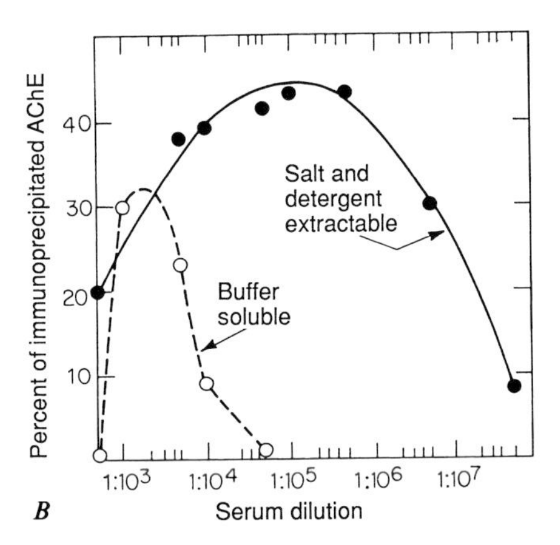

cells. Masson, ×75. Insets: Electromyography records of voluntary movements of a dorsal interosseous muscle in the same patient. Recording was performed using a DISA electromyograph on admission (top) and after 3 months (bottom), when muscle atrophy developed. Note the myasthenic pattern manifested by a decrease in the amplitude on recurrent muscle activity, and the appearance of a neuropathic pattern exhibited by incomplete interference with giant and polyphasic potentials, after 3 months. *B* preferential immunoreaction with membrane-bound muscle AChE. Buffer-soluble and salt- and detergent-extractable fractions of muscle homogenates were prepared as described previously [357] from the thigh muscle of an 18-week fetus. Enzyme inhibition by the patient serum was studied using direct immunoprecipitation. AChE activity was measured spectrophotometrically according to Ellman et al. [112]. Normal human serum served as a control. Values presented are the percentage inhibition of AChE, calculated as enzymatic activity left in each supernatant after treatment with patient serum compared to the normal serum control. Note that the patient's serum preferentially inhibits the salt- and detergent-extractable (membrane-associated) enzyme.

the serum could contain antibodies specific to the muscle membrane enzyme in addition to those with the general anti-ChE interaction.

It is not clear yet whether AChE was the primary target for the autoimmunological reaction in this patient, or whether the anti-AChE antibodies are anti-idiotypes to natural antibodies directed against OP insecticides. In both cases, however, these findings implied that antibodies against muscle membrane AChE may play a major role in eliciting neuromuscular dysfunction, neuropathy, and muscle atrophy. To further examine this possibility, cDNA clones coding for human BuChE [264] and AChE [318] should be used to synthesize in heterologous expression systems, large amounts of the authentic human proteins, which will be injected into animals to examine the possibility that AChE and/or BuChE may induce atypical neuromuscular disorders.

The autoimmunogenic properties of AChE imply that proteins sharing sequence homologies with it are also vulnerable to such pathological phenomena, particularly under disease states. This specifically refers to Tg, excess amounts of which are released into the circulation in hyperthyroid diseases. Both AChE and BuChE display considerable sequence homologies with Tg [264, 315]. The highest level of homology is within the region produced in our pEX bacterial expression system [104]. To examine whether autoimmune antibodies to ChE are induced in hyperthyroid patients, we therefore tested the interaction of immunoglobulins (Ig) from such patients with the recombinant BuChE polypeptides from this system as well as with the enzyme in situ, in fixed muscle fibers. For this purpose, a dot-blot immunoreaction with the pEX_3 BuChE cDNA-derived polypeptides was performed, using diluted antisera and iodinated protein A. Igs from 3 normal individuals demonstrated nil binding above background. The polyclonal rabbit anticloned ChE showed the highest level of binding as expected. Six of 9 patients showed evidence of significant binding to the recombinant protein [214].

Further analyses included immunoblots of the recombinant BuChE products following gel electrophoresis. Purified Tg served as a positive control in these experiments. Using these polyclonal antibodies, faint but distinct bands could be detected for Tg with recombinant BuChE and vice versa. Five normal Igs had undetectable binding to either BuChE or Tg. Hyperthyroid patients demonstrated binding to BuChE but this was not always accompanied by binding to Tg. In 3 patients with hyper-thyroidism there was no binding to BuChE despite high levels of binding to Tg.

Immunolocalization of the responsive target protein to the autoimmune antibodies in hyperthyroid patients was done using the in situ immunoreaction technique described in Section II.4.d. When muscle fibers were incubated in polyclonal antirecombinant BuChE serum or with Igs from 2 of the patients with Graves' ophthalmopathy, silver grains were apparent in the emulsion above areas of dense ChE activity at endplate regions [214]. Similar results were obtained with 6 GO Igs tested whilst 3 normal Igs did not bind to any part of the muscle fibres.

6. Cholinesterase Genes Are Expressed in the Haploid Genome Oocytes

Dot-blot hybridization of poly(A)$^+$ RNA from ovarian tissue samples from several individuals, using ^{32}P-labelled BuChE cDNA, indicated that the level of ChE mRNAs in the mature human ovary is $< 0.001\%$ of total mRNA [218]. However, reproducible in situ hybridization signals were observed in ovarian sections from various individuals and were localized in single cells, identified as oocytes, within follicular structures of different developmental stages (fig. 8) [316]. In a semiquantitative analysis performed on 71 primordial, 14 preantral and 20 antral follicles with positively labelled oocytes, the average grain density of hybridization signals on a scale of 1–5 was estimated to be 1–2, 4–5 and 2–4 respectively [219]. Thus the level of BuChE mRNA appeared to be reproducibly high in preantral follicles as compared with those which were primordial and antral, while atretic follicles remained negative and other cell types did not display significant labelling [219, 316]. No signals could be detected in many follicles, probably due to different sectioning levels, where many sections did not cut through the oocytes themselves. No labelling was observed in control experiments using an irrelevant DNA fragment from a non-expressed intron of the human superoxide dismutase gene [201] or following pretreatment of the ovarian sections with pancreatic ribonuclease. Over 550 grains could be counted in single oocytes exposed for 15 days, indicating high levels of ChE mRNA within the oocyte (fig. 9). For comparison, similar levels were detected under the same hybridization conditions for the pCO_2 mRNA, representing 15% of the total mRNA content in two-cell sea urchin embryos [76]. Considering the low content of oocytes (about 1×10^5 in a mature ovary) out of the total number of ovarian cells, this high intensity of labelling is compatible with the 0.001% levels of BuChE mRNA detected by dot-blot hybridization and suggests that BuChE is expressed in high levels in the human oocyte throughout its development, with a transient increase in the preantral phase.

Biochemical properties of the ovarian ChE were examined by sucrose gradient centrifugation followed by measuring [^{3}H]-ACh hydrolysis in the presence of selective inhibitors. In several such analyses, ovarian ChE activity was defined as 'true' AChE, sensitive to inhibition by 10^{-5} *M* of the selective anti-AChE inhibitor BW284C51 and resistant to the anti-

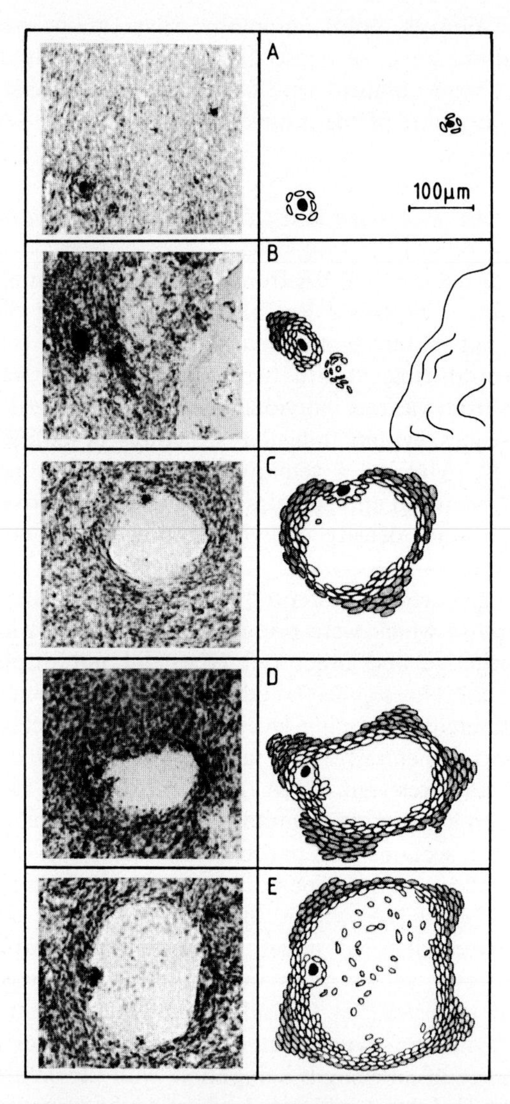

Fig. 8. Active transcription of BuChE mRNA in developing human oocytes revealed by in situ hybridization with ^{35}S-BuChE cDNA. Full-length purified BuChE cDNA [264], was labeled with (^{35}S)deoxyadenosine and deoxycytosine [123] to a specific activity of 3×10^8 cmp/μg and employed as a probe for hybridization in situ with frozen 12-μm sections from mature, normal ovaries. Freshly prepared sections were treated essentially according to Branks and Wilson [43] and as detailed elsewhere [219]. Exposure under Kodak NTB-2 emulsion was 12 days at 4 °C. Counterstaining was with hematoxylin and eosin. Photographs and drawings of follicles in primordial *(A)*, preantral *(B)* and antral stages *(C–E)* are displayed. Note absence of labeling in atretic follicle (*B*, right) and intense labeling in preantral follicle (*B*, left) [reproduced from 316, with permission].

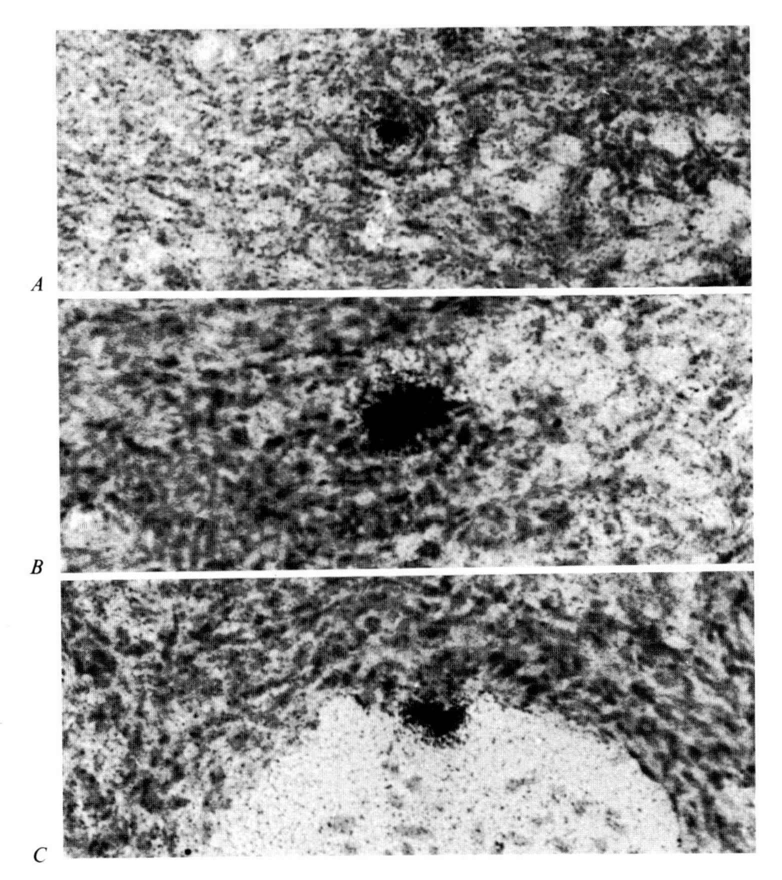

Fig. 9. BuChE mRNA levels in developing oocytes. In situ hybridization was as in the legend to figure 8. High magnification photographs are displayed for follicles at the primordial *(A)*, preantral *(B)* and antral *(C)* stages. Note relatively intense labeling of oocyte at the preantral stage [reproduced from 316, with permission].

BuChE inhibitor iso-OMPA [316]. However, careful inspection of cross-hybridization levels between AChE cDNA and BuChE cDNA revealed high specificity of the hybridization signals [191], strongly suggesting that the in situ hybridization of BuChE cDNA with ovarian sections selectively revealed the presence of BuChE mRNA. Further experiments using AChE cDNA demonstrated that AChE mRNA is also produced in the oocytes, although in lower quantities [Ayalon et al., unpubl. observations]. It should be noted that the presence of mRNA transcripts does not necessarily imply that their corresponding protein products are present and biologically active. This is particularly true in the case of oocytes, where accumulation of mRNA transcripts for later use after fertilization has been observed [313, 352]. Cytochemical staining experiments will have to be performed on ovarian sections to resolve this issue.

The ovarian AChE appeared to be predominantly soluble in low-ionic strength buffer, but up to 30% of the activity remained attached to the buffer-insoluble fraction of the ovarian homogenates, from which it could be removed by 0.1% of the nonionic detergent Triton X-100. The buffer-soluble AChE sedimented as 5.5–6.5 globular dimers, whereas the detergent-extractable enzyme presented the properties of both 4–5S monomers and heavier dimers, indicating that it could reflect contamination with erythrocytes. Minor activities of BuChE tetramers, most probably of plasma origin, could be observed in the buffer-soluble fractions of part of the samples and were completely resistant to BW284C51 but sensitive to iso-OMPA at the above concentrations. This most probably implies that the resolution of conservative biochemical methods is not sufficient for analysis of this cell type-specific phenomenon, which can only be resolved by combination of molecular biology and cytochemical approaches.

7. De novo Inheritable Amplification of the CHE Gene in a Family under Exposure to Parathion

Two loci related to inherited alterations in serum CHE [305, 347] were genetically linked with the transferrin [14, 161, 321] and haptoglobin [211] genes on chromosomes 3 and 16, respectively, but it remained unclear whether they represent structural CHE genes. We demonstrated by in situ chromosomal hybridization that both chromosome 3 and 16 carry sequences hybridizing with BuChE cDNA [317]. On chromosome 16, the hybridizing sequences localized to an area close to that where the haptoglobin gene was localized [306] and where the esterase B gene ESB3 [17] has been provisionally mapped. On chromosome 3, these sequences localized to a 3q site that is commonly abberated, and related to abnormal megakaryocyte proliferation, in acute myelodysplastic anomalies [262].

In situ hybridization experiments were performed using Q-banded and R-banded chromosome preparations [226] from peripheral blood lymphocytes and either a 760 nucleotide long (^{35}S) cDNA probe coding for about half of the catalytic subunit of BuChE isolated from fetal human brain [263] or a 2,230 nucleotide long EcoRl fragment derived from a full-length BuChE cDNA from both fetal and adult liver origin. Hybridization was carried out according to Rabin et al. [267] with some modifications.

Of a total of 52 cells from 8 unrelated volunteers having normal karyotypes which were scored, 53 copies of chromosome No. 3 in 43 cells and 37 copies of chromosome No. 16 in 30 cells gave positive hybridization signals. These carried 98 and 77 grains on chromosomes 3 and 16, respectively, altogether 175 grains out of a total of 646 which were associated with chromosomes, with 45 (87%) cells being positive for either one or both chromosomes [317].

The cumulative distribution of autoradiographic silver grains observed over the photographed chromosome spreads from normal individuals was plotted on a histogram representing the haploid human genome and divided into equal units scaled to the average diameter of a silver grain (0.35 μm). This analysis revealed that 63 (64%) of the grains on chromosome 3 were concentrated within the region 3q21 → 3qter, with two clear peaks around 3q21 and 3q26-ter. On the shorter chromosome 16, radioactivity concentrated around the 16q12 band, with 49 (64%) of the grains within the region 16q11 → 16q23. Statistical evaluation of the number of silver grains per unit chromosome length, assuming a Poisson distribution, indicated that the localization on chromosomes 3 and 16 was significant in both cases ($p < 0.025$ and $p < 0.01$, respectively). The labelling over all other chromosomes was not significant. Both probes from the BuChE cDNA clone gave essentially similar results, further confirming the significance of the above-mentioned hybridization experiments.

The gene mapping approach was further employed to detect the in loco amplification of a defective CHE gene in a family of farmers designated the H family. The occurrence of an unusual BuChE phenotype in the H family was brought to our attention when one of its members, I.T., suffered from characteristic prolonged apnea [158] following a single intravenous administration of succinylcholine during the course of general anesthesia. On another occasion, a sibling, M.I., fainted while spraying parathion in a cotton field. Examination of serum BuChE activity in all members of this family revealed very low levels of BuSCh-hydrolyzing activity in serum samples from both I.T., and M.I., with increased sensitivity to the BuChE-specific OP inhibitor iso-OMPA [21] and pronounced resistance to the local anesthetic dibucaine (2-butoxy-N-(2-diethylaminoethyl)-4-quinoline carboxamide), all being characteristic of an atypical BuChE enzyme [348]. Figure 10 and table 1 present these results.

To examine whether the expression of the unusual BuChE phenotype was due to alteration(s) at the level of the CHE genes, DNA blot hybridization experiments were performed, using 32 P-labeled fragments from the cloned BuChE cDNA as probes. When digested with the enzymes EcoRI and Hind III and probed with full-length BuChE cDNA, peripheral blood DNA from M.I. revealed highly positive restriction fragments of approximately 6.0 kb and smaller (for EcoRI) and 2.5 kb (for Hind III). These were absent from DNA from several other members of the H family,

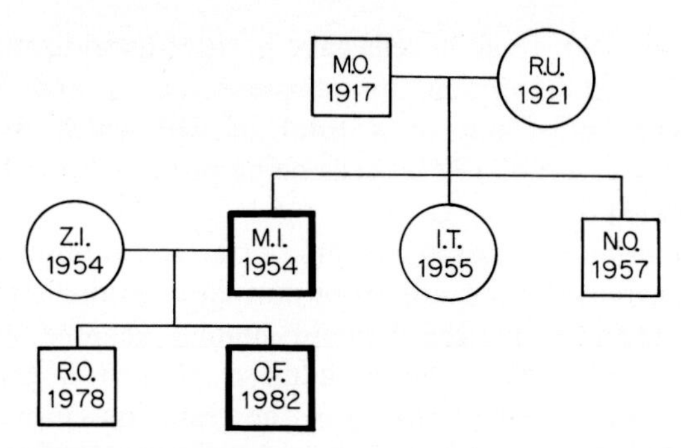

Fig. 10. The H family pedigree and birth years. Individuals with amplified CHE genes are presented by bold boxes.

Table 1. Serum BuChe activities in members of H family

No.	BuChE activity[1]	Dibucaine IC_{50}[2]	Iso-OMPA inhibition[3]
1 R.U.	5.2 ± 0.6	1.0	31
2 M.O.	4.3 ± 0.5	7.5	31
3 N.O.	3.5 ± 0.4	10.0	45
4 I.T.	1.1 ± 0.2	50.0	78
5 M.I.	0.8 ± 0.1	100.0	82
6 Z.I.	5.1 ± 0.2	10.0	36
7 R.O.	4.9 ± 0.3	10.0	31
8 O.F.	2.1 ± 0.4	10.0	44
9 Control[4]	6.6 ± 1.3	6.3	38

[1] BuChE activity, in μmol thiocholine released/ml serum/min, was determined spectrophotometrically by the acetylthiocholine hydrolysis technique [112]. Assays were performed in multiwell plates and several time points were measured in a Bio-Tek EL-309 microplate reader. Spontaneous release of thiocholine was subtracted and rates of enzymatic activity were calculated from linear regression curves of optical density at 405 nm. Values are presented as mean ± SEM (standard evaluation of the mean) from three independent measurements of three different serum samples for each person. Time intervals between samplings were at least 6 months.

[2] Dibucaine inhibition is expressed as the micromolar concentration of inhibitor that blocks 50% of thiocholine release under the assay conditions detailed in footnote 1.

[3] The percentage of thiocholine release that was inhibited by increasing the concentration of the OP inhibitor iso-OMPA [21] from 0.1 μ*M* to 0.01 m*M* is presented.

[4] Control serum was a pooled mixture of equal volumes of serum samples from 10 apparently healthy individuals and with normal BuChE. Note the high level of thiocholine release, the low IC_{50} values for dibucaine and the moderate iso-OMPA inhibition for control serum as compared with samples 4 and 5 [reproduced from 265].

specifically including the parents, R.U. and M.O., and the sibling, I.T. [265]. This pattern was reproducibly obtained using DNA samples taken at long (6 months) time intervals. Interestingly, the hybridization signal with these amplified fragments was considerably weaker when the 5′ and 3′ terminal parts of BuChE cDNA were used as probes, suggesting that the initial amplification event was confined mainly to the central part of the CHE gene and that the external parts of this gene were amplified to a lesser extent, in agreement with the 'onion skin' model for other amplification units as opposed to a pattern of tandem repeats [325]. Furthermore, the variable sizes of the positive fragments obtained with EcoRI but not with Hind III, also argue against the tandem repeat model for the initial event of the amplification (fig. 11).

To quantitate this amplification, a dot-blot hybridization was performed with six dilutions of peripheral blood DNA from each member of the H family. DNA from M.I. and one of his sons, O.F., contained an equivalent of about 25 pg of BuChE cDNA-positive sequences per microgram of genomic DNA, whereas DNA from other members of the family displayed levels equivalent to only 1–3 pg/μg DNA (fig. 12). Blot hybridization of DNA from O.F. revealed restriction patterns similar to those of M.I. Altogether, these hybridization experiments suggest that the de novo amplification event occurred in the genome of M.I. very early in spermatogenesis, embryogenesis, or oogenesis, rendering it inheritable. However, we cannot rule out the possibility that O.F. inherited the predisposition for this amplification and that the occurrence of the amplified DNA in his case was due to the prolonged exposure to parathion.

When metaphase peripheral blood chromosomes from M.I. and his mother, R.U., were analyzed by the Giemsa (G) and the bromodeoxyuracil-induced (R)-banding techniques [317]; apparently normal karyotypes were observed in both individuals, with neither minute chromosomes nor homogeneously staining regions (HSR) that are commonly found in cases of gene amplifications [293]. In situ hybridization with (^{35}S)-BuChE cDNA revealed intense labeling of the 3q29 region in M.I.'s chromosomes as compared with controls (fig. 13a–c). (^{35}S)-labeling was mostly confined to chromosomal structures, excluding the possibility that the amplified CHE genes were present in submicroscopic extrachromosomal elements.

Altogether, these observations suggested that the inheritable amplified DNA segment was localized close to the original site of the CHE gene at 3q26 [317]. Since the amplification unit corresponds primarily to the middle third of the cDNA, and assuming that it is present at a single site in the genome, we calculate that at least 100 copies of the amplified fragment are present in a possibly fully inheritable form in the genomic DNA of M.I. In

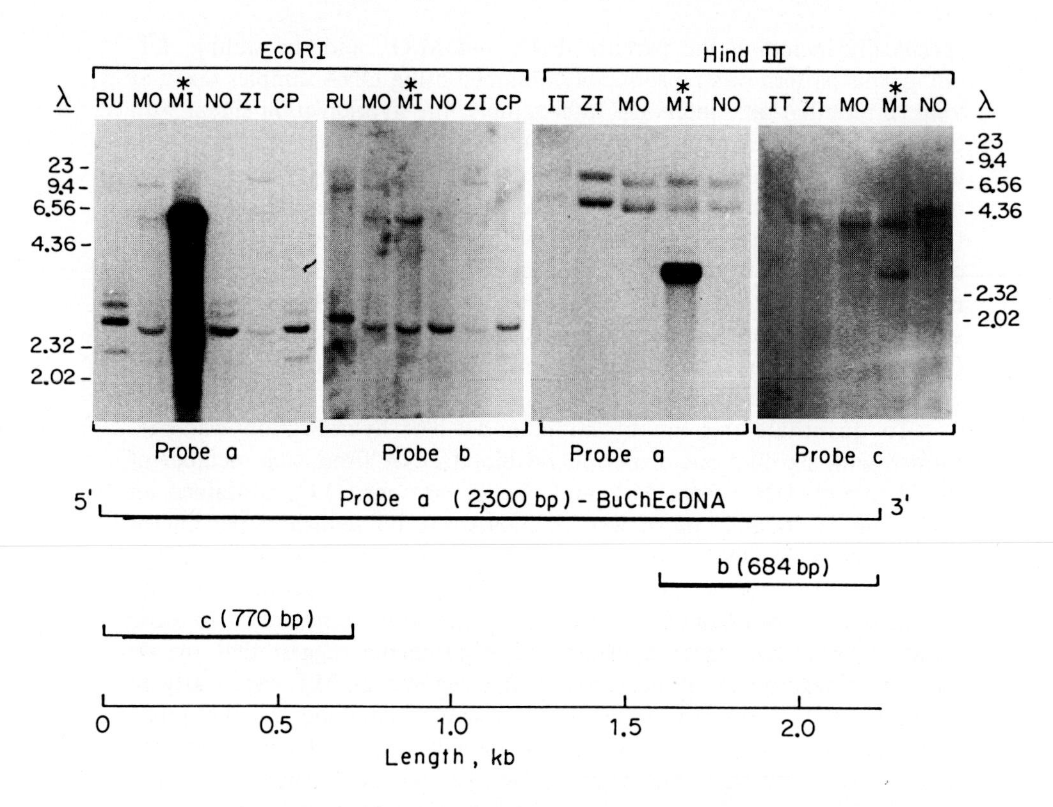

Fig. 11. DNA blot hybridization with regional BuChE cDNA probes. 10-μg samples of DNA from peripheral blood were digested to completion with excess amounts of the restriction enzymes EcoRI (Pharmacia) or Hind III (IBI), electrophoresed on 1.0% agarose gel and transferred to a Gene-Screen filter (NEN). Subcloned BuChE cDNA probes a, b and c were prepared as described [264, 315] and labeled with [^{32}P]-dATP by the random primer technique [123]. Hybridization was carried out for 16 h and filters were washed in 0.15 *M* NaCl/0.015 *M* sodium citrate/0.1% sodium dodecyl sulfate at 60 °C. Exposure was for 6 days at −70 °C with an intensifying screen. Lambda phage DNA digested with the restriction enzyme, Hind III served for size markers. The lanes loaded with DNA from M.I., the individual who suffered from parathion poisoning, are marked with an asterisk. The probes used are schematically drawn below, with coding regions represented by thick lines and untranslated regions represented by thinner lines. Note that the labeled cDNA probes light up the prominent 2.0- to 2.5-kb fragments in genomic DNA from the parents, siblings, and spouse of M.I., in agreement with previous results [264, 315]. The relative weak 4.7- and 9.4-kb genomic DNA fragments that also hybridize with this probe [264, 315] can hardly be seen under these exposure conditions. In contrast, DNA from M.I. reveals strong hybridization signals with EcoRI-cut fragments of variable sizes. bp = Base pairs [reproduced from 265].

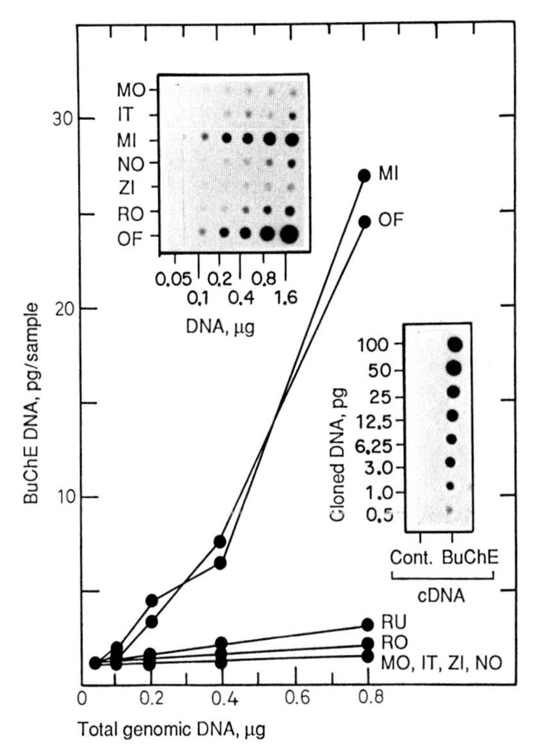

Fig. 12. Quantification of amplified CHE genes in members of H family by dot-blot hybridization. Denatured genomic DNA from peripheral blood cells was spotted onto a GeneScreen filter (New England Nuclear) using a dot-blot applicator (BioRad). Electroeluted BuChE cDNA (probe a in figure 11) was spotted in parallel for calibration. All samples contained the noted quantities of genomic DNA and denatured herring testes DNA to yield a total amount of 2 µg of DNA per spot. Hybridization, wash and exposure were done with (^{32}P)-labeled probe a, as in the legend of figure 11. Quantities of genomic BuChE cDNA sequences that hybridized with the labeled probe in each member of the H family were determined in values equivalent to picograms of BuChE cDNA by optical densitometry of the exposed X-ray film (Agfa-Gevaert) in a Bio-Tek microplate reader. An irrelevant clone served as control. Insets: Autoradiographed films showing dot hybridizations of micrograms of genomic DNA (upper) and picograms of cloned DNA (lower). Cont = control [reproduced from 265].

spite of the apparent gene amplification in M.I. and O.F., gel electrophoresis and immunoblot analysis of serum proteins with anti-BuChE antibodies failed to reveal overexpression of serum BuChE in these individuals [105]. However, this does not exclude the possibility that the amplified gene was overexpressed early in development, for example in germline cells and embryonic tissues, where the CHE gene is intensely expressed [196, 316].

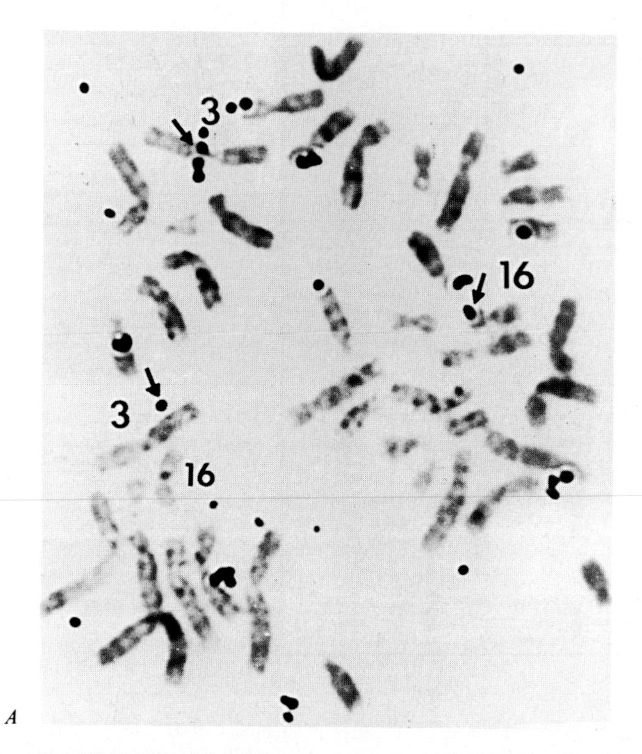

A

Fig. 13. A Normal lymphocyte metaphase chromosome spread after hybridization with (^{35}S)-BuChE cDNA and autoradiography (exposure: 12 days). A representative R-banded partial spread, hybridized with probe A, is displayed, in which both chromosomes 3 and one chromosome 16 are labeled in their q arms [reproduced from 317, with permission]. *B* Chromosomal mapping of human ChE genes by in situ hybridization. Distribution of silver grains scored over human chromosomes from 52 Q- or R-banded metaphase spreads is presented for 31 and 21 metaphase spreads, respectively, hybridized with (^{35}S)-BuChE cDNA probes. Results of experiments carried out with full-length or partial probes were essentially similar. A high concentration of grains was located on the chromosome regions 3q21–q26 and 16p11–16q23, while labeling on all other chromosomes was insignificant [reproduced from 317, with permission]. *C* Chromosomal mapping of amplified CHE gene by in situ hybridization. Lymphocyte metaphase chromosomes from M.I. (see fig. 10–12) were employed for in situ hybridization with (^{35}S)-labeled probe as described [317]. Exposure was for 2–4 days under NTB-2 emulsion (Kodak) diluted 1:1 with H_2O, and development was for 30 s with Kodak HC-110 developer. Four representative R-banded partial spreads are displayed, in all of which the q arm of chromosome No. 3 is labeled (see arrows). Because of the three-dimensional structure of the chromosomes covered with the photographic emulsion, disintegrations and grain formation occur more frequently close to the chromosome than directly on it, where the emulsion layer tends to be thinner (see fig. 13A). Comparative mapping of amplified and normal CHE genes on chromosomes 3 and 16 was performed by plotting the distribution of silver grains scored over chromosomes No. 3 and 16 from 21 R-banded metaphase

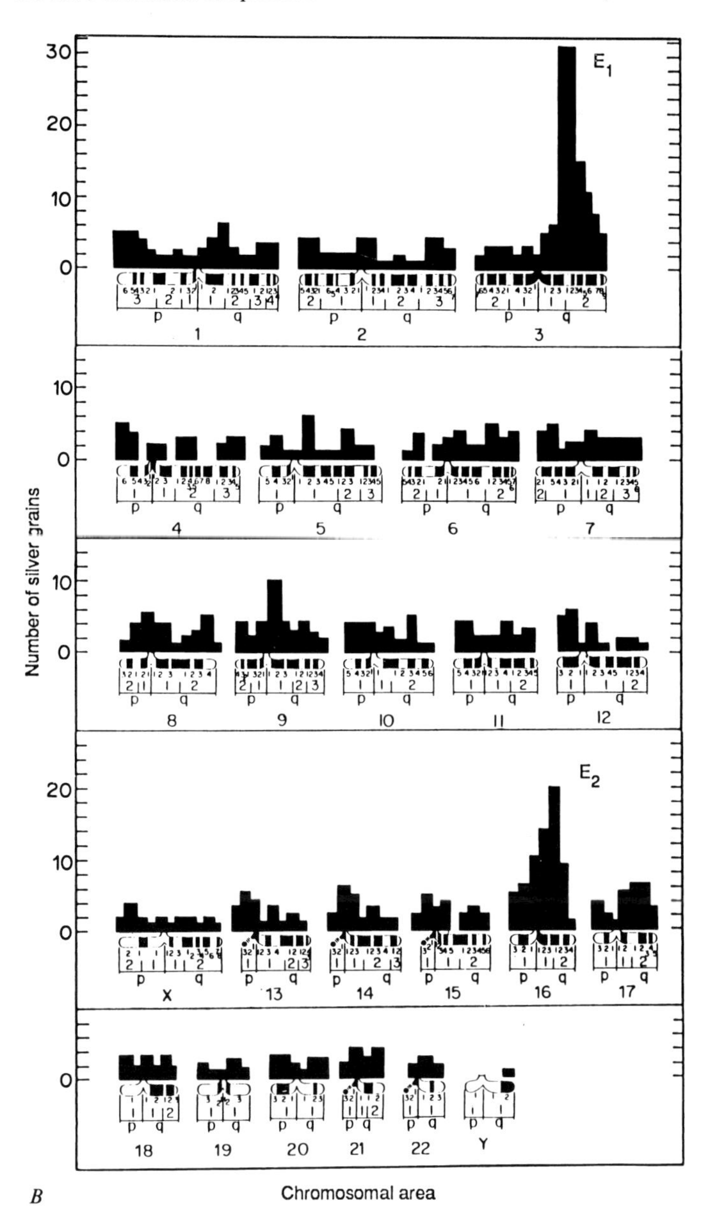

B

spreads prepared from R.U. and M.I.'s peripheral blood. This analysis revealed that relatively high percentage of grains was located on the 3q27ter region with M.I.'s chromosomes, with a prominent peak on 3q29 (see fig. 13D, p. 44). Parallel mapping using R.U.'s chromosomes revealed a main peak on 3q26, similar to our previous mapping (Cont.) [317]. Labeling on chromosome 16 was essentially similar in M.I., R.U. and control [reproduced from 265].

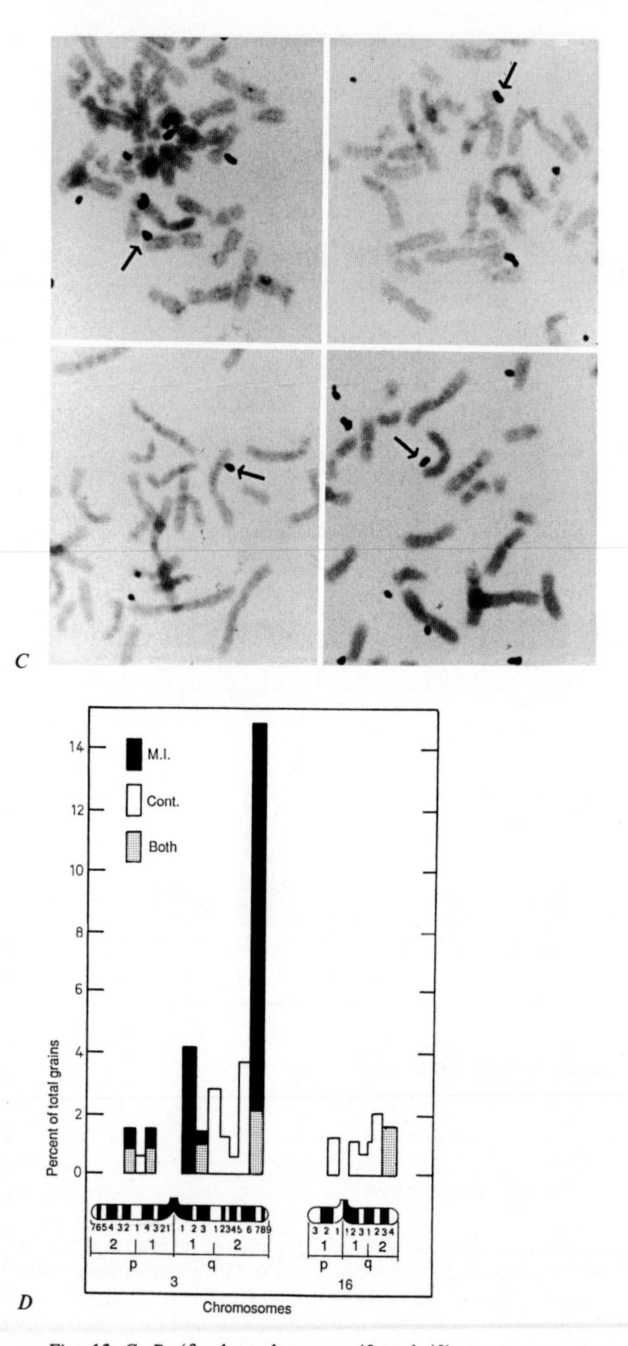

Fig. 13. C, D. (for legend see pp. 42 and 43)

8. Coamplification of Human Acetylcholinesterase and Butyrylcholinesterase Genes in Blood Cells

Cholinesterases are expressed in leukemic cells, where they are known to be subject to differentiation-related modulations [33]. In order to search for putative structural changes within the human ACHE and CHE genes encoding AChE and BuChE in leukemic DNA samples, their restriction fragment patterns were examined in peripheral blood DNA from 16 patients with various leukemias as compared with DNA from 30 healthy individuals. For this purpose, DNA blot hybridization was performed with equal amounts of patients' DNA following complete digestion with the restriction endonucleases PvuII and EcoRI and gel electrophoresis. Hybridization with 32P-labeled AChE cDNA and BuChE cDNA repeatedly revealed invariant restriction patterns and signal intensities for DNA from all of the healthy individuals. The same restriction pattern and signal intensities were observed in DNA from 12 of the leukemic patients. In contrast, the hybridization patterns in the 4 remaining samples displayed both qualitative alterations and a clear signal enhancement with both cDNA probes [191]. It revealed intensified labeling of bands that also existed in the control lane, as well as the appearance of various novel labeled bands.

The relationships between the high levels of AChE in megakaryocytes [56] on one hand and blood platelet and erythrocyte formations on the other hand [74] imply a correlation of ChEs to hemocytopoiesis. In view of previous reports correlating ChEs with megakaryocytopoiesis and platelet production [54, 55, 258], DNA from additional patients with platelet disorders, whether or not defined as leukemic, was similarly examined. Significantly enhanced hybridization signals with both cDNA probes were found in 8 out of 10 such patients examined, 1 of them leukemic. Interestingly, the intensity of hybridization in several of these samples was much higher than it was in any of the previously tested leukemic DNA samples. Furthermore, the amplification events in two of these samples appeared to involve many additional PvuII-cut DNA fragments, due to either nucleotide changes producing novel PvuII restriction sites, or different regions of DNA having been amplified [191].

To further compare the restriction fragment patterns of the amplified genes, the relevant lanes from these autoradiograms were subjected to optical densitometry. This analysis clearly demonstrated the appearance of slightly enhanced hybridization signals at equal migration positions to those observed in control DNA for a representative leukemic DNA sample with a moderate amplification. In another leukemic DNA sample taken from a patient with reduced platelet counts, the densitometry signals were higher by an order of magnitude and presented several additional short

PvuII-cut fragments. Yet much higher signals and more novel bands of various sizes were observed with a third sample, derived from a nonleukemic patient with a pronounced decrease in platelet counts (thrombocytopenia). Figure 14 presents these findings.

The variable degrees of amplification occurring in the genes coding for AChE and BuChE in these individuals were quantified by slot-blot DNA hybridization, using a 5-fold dilution pattern. Cross-hybridization between the two cDNA probes was exceedingly low (<0.01), demonstrating that the observed amplification events occurred in each of these genes separately and did not merely reflect similarity in their sequences. One microgram of genomic DNA from the patients with CHE and ACHE gene amplifications included genomic sequences equivalent to about 0.1 and 0.01 ng of the purified BuChE cDNA and AChE cDNA inserts, respectively. Parallel analysis using similar quantities of control DNA revealed considerably lower signals with both probes (fig. 15).

Taking the total complexity of human genomic DNA as $4 \cdot 10^{-9}$ base pairs, this implied that at least 40–100 copies of both these sequences are present in the above DNA. Other examined DNAs featured about 10-fold lower signals with BuChE cDNA, reflecting a more modest amplification in an order of up to 20 copies per genome [191]. Repeated hybridization of

---

Fig. 14. Intensified amplification is accompanied by structural differences within the amplified DNA regions. Comparative analysis of representative DNA samples from a healthy control (C1), a leukemic AML patient with moderate amplification (L04), and a nonleukemic patient with a pronounced decrease in platelet counts (P03) was performed by DNA blot hybridization using ^{32}P-labeled probes. *A* Blot hybridization with PvuII-cut genomic DNA and AChE cDNA probe (Ac) and with EcoRI-cut genomic DNA and BuChE cDNA probe (Bt). *B* Optical densitometry of individual lanes from the PvuII-treated AChE cDNA-hybridized blot was performed at 545 nm [265]. Note the increased intensity of the densitometric measurements and the appearance of additional labeled restriction fragments in P03 and L04 lanes as compared with L70 and C1. *C* Restriction sites for PvuII and EcoRI on the cDNA probes. Note that the number of PvuII-cut DNA fragments in P03 that were labeled with AChE cDNA exceeds the expected number of three fragments based on the PvuII restriction pattern of AChE cDNA, reflecting structural changes and appearance of PvuII restriction sites within the amplified DNA sequence. Exposure was for 6 days [Reproduced from 191].

Fig. 15. Quantification of amplification levels in diseased DNA samples by slot-blot hybridization. Denatured genomic P03 DNA, analyzed in figure 14, was spotted onto a Gene-Screen filter using a slot-blot applicator (BioRad). Electroeluted AChE cDNA (Ac) and BuChE cDNA (Bt) (see fig. 14C) were spotted in parallel for calibration. Genomic DNA from an apparently healthy individual (C2) and herring testes DNA (Co) served as controls. All samples contained the noted quantities of genomic or insert DNAs supplemented with denatured herring testes DNA to yield a total of 2 μg of DNA per slot. Hybridization, wash, and exposure were done as detailed elsewhere [265] with ^{32}P-labeled AChE cDNA or BuChE cDNA as noted. Note the minimal levels of cross-labeling between the corresponding cDNAs and the intense labeling of P03 DNA as compared with controls [reproduced from 191].

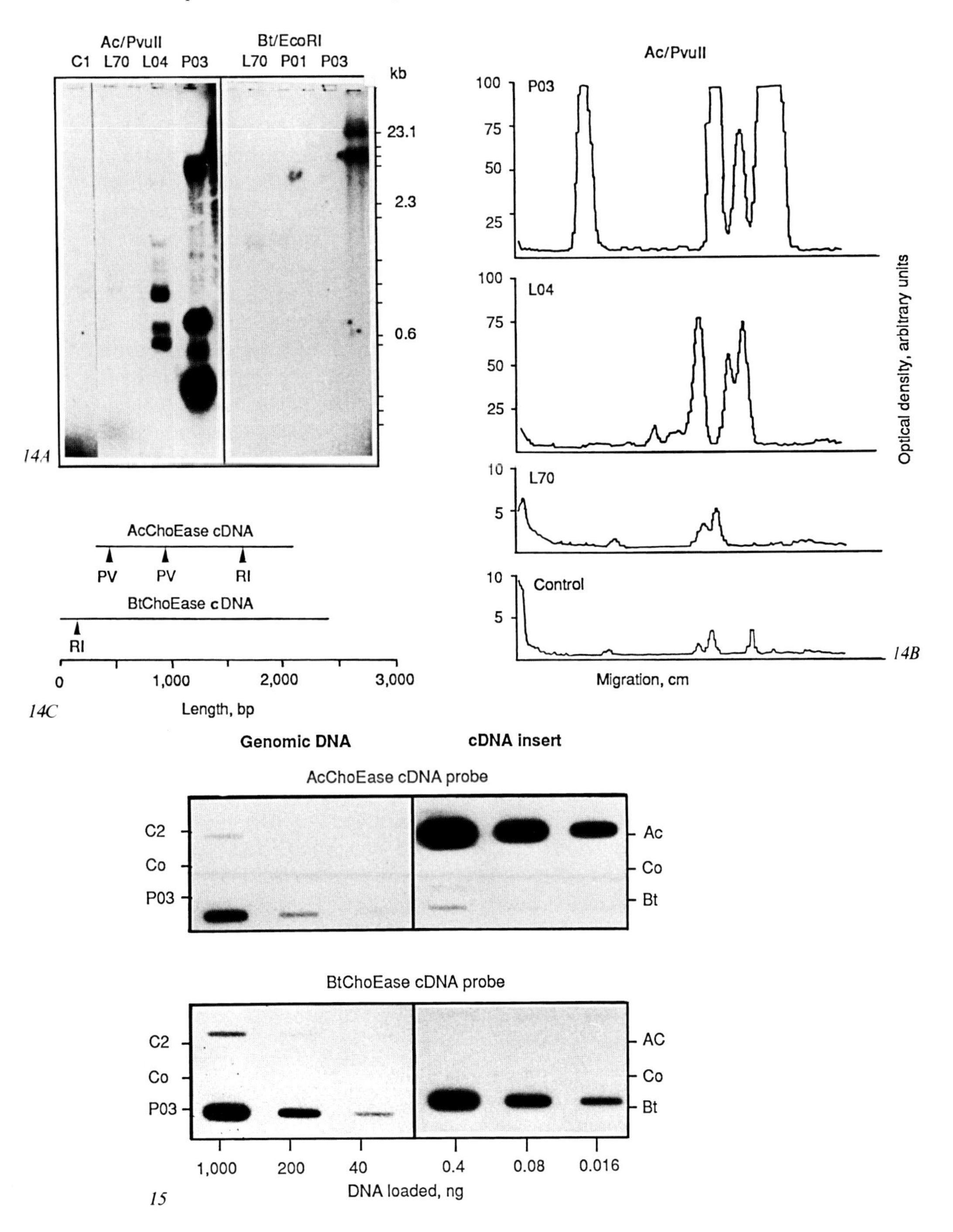

Ac/PvuII
C1 L70 L04 P03
Bt/EcoRI
L70 P01 P03
kb
23.1
2.3
0.6
14A
Ac/PvuII
P03
L04
L70
Control
Optical density, arbitrary units
Migration, cm
14B
AcChoEase cDNA
PV PV RI
BtChoEase cDNA
RI
0 1,000 2,000 3,000
Length, bp
14C
Genomic DNA
cDNA insert
AcChoEase cDNA probe
BtChoEase cDNA probe
C2
Co
P03
Ac
Co
Bt
AC
1,000 200 40 0.4 0.08 0.016
DNA loaded, ng
15

the same blots with a cDNA probe coding for the nonrelated rat ribosomal protein L19, demonstrated no amplification in all of the examined samples and similar labeling intensities for both patient and control DNAs.

Altogether, 11 cases of coamplification within the ACHE and CHE genes were observed in DNA samples from 20 patients with abnormal hematocytopoiesis, while DNA from 30 healthy individuals showed neither amplification nor polymorphism with respect to the restriction patterns obtained with these probes. The DNA samples presenting 6 of these amplifications were derived from 4 cases of AML with 10–50 copies of both ACHE and CHE genes, and 3 cases of platelet count abnormalities. One expressed excess platelet counts and 10–20 copies of the ACHE and CHE genes, and the two others showed reduction of platelet counts and featured 10–200 copies of the same genes. These striking concomitant multiplications presented a highly significant correlation ($p < 0.001$) between amplification of CHE-encoding genes and the occurrence of abnormal myeloid progenitor cells or promegakaryocytes in the examined individuals (table 2).

9. Altered Expression of Cholinesterase Genes in Carcinoma Patients under Antitumor Therapy

Several earlier reports have noted changes in the expression of ChEs in cases of carcinoma tumors [100, 347]. This has mostly been referred to as representing the expression of yet another 'embryonic' gene in neoplasia [73]. To find out whether carcinomas in general are accompanied by changes in the biochemical properties of serum ChE, we set out to determine inhibition curves for the ChEs in 77 serum samples drawn from patients suffering from carcinomas of various tissue origins and in 21 serum samples from healthy volunteers, using selective inhibitors specific for particular types of ChEs. Three compounds were used: iso-OMPA, an OP poison with high specificity towards serum BuChE [21]; BW284C51, a bisquaternary reversible inhibitor of 'true' AChE, and succinylcholine, a substrate analog of ACh which is rapidly hydrolyzed by normal BuChE but cannot be degraded by the 'atypic' or 'silent' types of serum BuChE [347]. Each inhibitor was added at six different dilutions, covering a wide range of concentrations.

The enzyme in all of the serum samples was clearly sensitive to iso-OMPA, as expected from human BuChE, with 1×10^{-6} *M* up to 1×10^{-4} *M* of the inhibitor sufficient for quantitative block of ACh hydrolysis. The inhibition curves of the tumor serum samples could be divided into three major groups: (1) curves which are indistinguishable

Table 2. Appearance of amplified ACHE and CHE genes in hematocytopoietic disorders

No.	Type	Defective progenitors	Approx. amplification	
			ACHE	CHE
Leukemias[1]				
1(L23)	AML	myeloid	N	N
2(L38)	AMegL	promegakaryocytes	N	N
3(L26)	AMOL	monocytes	N	N
4(L10)	AML	myeloid	5–15	5–15
5(L41)	AMML	myeloid/monocytes	N	N
6(L42)	AML	myeloid	N	N
7(L79)	AML	myeloid	N	N
8(L70)	AML	myeloid	10–20	10–20
9(L20)	AML	myeloid	N	N
10(L96)	AML	myeloid	N	N
11(L62)	AMML	myeloid/monocytes	10–20	5–15
12(L59)	AMML	myeloid/monocytes	N	N
13(L15)	AML	myeloid	N	N
14(L12)	AML	myeloid	N	N
15(L03)	AMLM2	myeloid	N	N
16(L04)	AMLM2	myeloid	40–60	10–20
Megakaryocytopoietic disorders[2]				
	platelet count			
16(L04)	low	promegakaryocytes	40–60	10–20
17(PO1)	high	promegakaryocytes	10–20	10–20
18(PO2)	low	promegakaryocytes	N	N
19(PO3)	low	promegakaryocytes	40–100	50–200
20(PO4)	low	promegakaryocytes	N	N
Controls[3]				
21(C1)	normal	none	N	N
22(C2)	normal	none	N	N
23–50(Cx)	normal	none	N	N

The characteristic types of hematopoietic progenitor cells, which appear to be defective in each class of the screened leukemias, are noted [262]. The approximate extent of amplification was separately determined for ACHE and CHE genes by slot-blot DNA hybridization and optical densitometry. Numbers reflect the -fold increase in the calculated number of copies as compared with control DNA. N = normal.

[1] Peripheral blood DNA from 14 leukemic patients was received, together with clinical classification of the disease type, from *E. canaani* (The Weizmann Institute of Science). Two other patients (LO3 and LO4) were diagnosed and classified in our department (HZ). (AMegL = acute megakaryocytic leukemia; AMOL = acute monocytic leukemia; AMML = acute monocytic/myeloid leukemia; AMLM2 = Fab subclassification of AML).

[2] Peripheral blood DNA from 5 patients from our department (HZ), suffering from abnormal platelet counts, was analyzed as detailed above. Abnormalities in platelet counts are noted, where 'low' implies $<80{,}000/mm^3$ and 'high' is $>600{,}000/mm^3$ (normal counts are considered $150{,}000-400{,}000$ platelets/mm^3). Note that LO4 (No. 16) appears twice.

[3] DNA samples from currently healthy individuals with normal platelet counts and blood ChE activities served as controls and were analyzed as detailed above, C1 and C2 correspond to representative control DNAs. Similar results were obtained in 28 more controls (Cx) [reproduced from 191].

Table 3. Clinical and biochemical characterization of carcinomas serum samples

No.	Age/sex	Ethnic origin	Type of tumor		Clinical treatment				Protein conc. in serum mg/ml	ChE sp. act. in serum pmol/μg P/h	Percent inhibition of ChE activity		
			histologic type	metastases	Surg	Irrad	Chem	Horm			1.10^{-4} *M* iso-OMPA	1.10^{-5} *M* BWZ84C51	50 ng/ml Suc Ch
Breast (100% females)													
1	66	Eur	Inf Duct	+	+	−	+	−	59	733	95	44	39
2	60	No Afr	Inf Duct	+	+	−	+	+	65	400	59	−18	6
3	68	Eur	Bil Comedu Ca	+	+	+	+	−	68	1,066	67	13	5
4	70	Eur	Inf Duct	−	+	−	+	−	62	600	77	9	61
5	69	Eur	Inf Duct	−	+	+	−	−	80	800	91	79	50
6	49	Asia	Inf Duct	−	+	+	+	−	64	666	79	4	46
7	41	Asia	Bil Adeno Ca	+	−	+	+	−	62	866	88	40	67
8	46	No Afr	Inf Duct	−	−	−	+	−	67	1,000	100	40	14
9	71	No Afr	Inf Duct	−	+	+	−	+	65	1,200	91	46	16
10	30	Asia	Inf Duct	−	−	+	+	−	70	666	89	8	20
11	70	Eur	Inf Duct	−	+	−	+	+	80	866	91	54	74
12	38	Eur	Inf Duct	−	+	+	+	−	68	1,000	91	5.7	91
13	63	Asia	In Op Ca	+	−	+	+	−	89	200	81	−11	92
14	56	Eur	Inf Duct	−	+	−	+	+	67	1,000	91	51	34
15	58	Eur	Inf Duct	−	+	+	+	+	67	1,000	85	13	96
16	47	Asia	Schirrous Ca	−	+	−	+	−	89	200	94	18	48
17	60	Eur	Medullary Ca	−	+	+	+	+	85	333	95	7.5	5
18	57	Eur	Inf Duct	−	+	−	+	−	77	933	93	32	50
19	68	Eur	Inf Duct	−	−	−	−	+	62	600	80	53	26
20	47	Asia	Inf Duct	−	+	−	+	+	65	466	100	−13	92
21	61	Eur	In Op Inf Duct	+	−	+	−	−	75	400	63	−5	22
22	53	Asia	Inf Duct	−	+	−	−	−	80	466	19	11	19
23	47	Eur	Inf Duct	−	+	−	+	−	77	666	100	27	56
24	69	Asia	Inf Duct	−	−	+	−	−	77	266	90	22	51
25	59	Eur	Medullary Ca	−	+	+	+	−	76	400	30	−9.2	98
26	30	Eur	Inf Duct	−	+	+	+	−	70	1,200	84	−2	43
27	40	Asia	Inf Duct	−	+	+	−	−	80	1,536	82	55	96
28	43	Asia	Inf Duct	−	+	+	−	−	78	533	84	14	14

Lung (male to female ratio 5/2)													
1	F/82	Eur	Oat cell	+	−	+	−	−	65	866	95	48	35
2	F/71	Eur	Sq cell well Dif Ca	−	+	+	−	−	66	600	96	13	102
3	M/53	Asia	Anaplastic Ca	+	−	+	+	−	65	333	88	40	48
4	M/61	Eur	Oat cell Ca	−	+	−	+	−	77	1,000	83	43	52
5	M/71	Asia	Sq cell Ca	−	+	+	−	−	70	533	94	6.5	63
6	M/46	Asia	Sq cell Ca	+	−	+	+	−	80	600	75	25	40
7	M/43	Eur	Adeno Ca	−	+	−	−	−	79	733	84	39	43
Digestive tract (male to female ratio 9/8)													
1	M/42	Eur	Adeno Ca of stomach	−	−	+	+	−	62	1,000	87	46	30
2	F/72	Eur	Adeno Ca colon	+	+	−	+	−	65	1,266	71	65	41
3	M/54	Asia	Adeno Ca colon	−	+	−	+	−	62	1,066	86	27	40
4	M/71	Eur	Adeno Ca colon	+	+	+	+	−	60	866	75	24	51
5	F/67	Eur	Adeno Ca colon	+	+	−	−	−	77	400	80	11	6
6	F/65	Eur	Adeno Ca rectum	−	−	−	+	−	74	733	90	7.7	17
7	M/73	Asia	Adeno Ca stomach	−	+	−	+	−	75	600	87	33	21
8	M/75	Eur	Adeno Ca rectum	−	+	−	+	−	61	400	73	40	4
9	F/71	Eur	Adeno Ca sigma	+	+	+	+	−	68	1,333	95.5	45.5	42
10	M/54	Eur	Adeno Ca colon	−	+	+	+	−	77	1,000	100	45	59
11	M/67	Eur	Adeno Ca stomach	−	+	+	−	−	77	666	100	3.3	6.5
12	M/67	Asia	Adeno Ca colon	−	+	+	+	−	68	533	85	0	36
13	F/60	Eur	Adeno Ca sigma	−	+	+	+	−	76	1,200	85	49	36
14	F/76	Eur	Adeno Ca colon	−	−	−	−	−	59	933	91	63	100
15	F/63	Eur	Adeno Ca stomach	−	−	−	−	−	40	1,133	95	−18	18
16	F/86	Asia	Adeno Ca stomach	−	+	−	−	−	45	666	100	43	30
17	F/80	Eur	Adeno Ca colon	−	−	−	−	−	63	466	100	34	53
Urinary tract (male to female ratio 4/1)													
1	M/72	Eur	Adeno Ca pros	+	+	+	+	−	75	1,066	88	24	5
2	M/82	Eur	Adeno Ca pros well Dif	−	−	+	−	−	80	800	60	78	45
3	M/86	Eur	Adeno Ca pros	−	+	−	−	+	65	1,066	84	16	38
4	M/76	Eur	Adeno Ca pros	−	+	−	−	+	67	733	80	23	50
5	F/70	Eur	Clear Cell Ca Kid	+	+	−	+	−	61	600	70	14	20
6	M/70	Eur	Adeno Ca pros	−	+	+	+	−	70	266	63	18	0
7	M/70	Eur	Clear Cell Ca Kid	+	+	−	−	−	75	600	75	18	69
8	F/70	Eur	Squamous Cell Ca bladder	−	+	−	−	−	72	533	85	23	9
9	M/68	Eur	Adeno Ca pros	+	−	+	+	−	76	1,200	88	5	22
10	M/72	No Afr	Adeno Ca pros	+	+	+	−	−	67	1,133	100	50	13

(continued on next page)

Table 3. (continued)

No.	Age/sex	Ethnic origin	Type of tumor		Clinical treatment				Protein conc. in serum mg/ml	ChE sp. act. in serum pmol/μg P/h	Percent inhibition of ChE activity		
			histologic type	metastases	Surg	Irrad	Chem	Horm			1.10^{-4} *M* iso-OMPA	1.10^{-5} *M* BWZ84C51	50 ng/ml Suc Ch
Gynecologic													
1	62	Asia	Squamous Cell Cx	+	–	+	+	+	64	733	90	38	47
2	55	Asia	Epidermoid Ca Cx	–	–	+	+	–	81	400	100	15	42
3	62	No Afr	Cyst Adeno Ca ovary	–	+	+	+	–	76	733	92	25	30
4	64	Eur	Pap Adeno Ca ovary	–	+	+	+	–	70	1,000	70	15	13
5	69	Eur	Pap Adeno Ca ovary	–	+	–	+	+	67	666	82	–30	16
6	62	Eur	Adeno Ca ovary	+	+	+	+	+	73	800	73	7	79
7	48	Asia	Adeno Ca ovary	–	+	+	+	–	80	1,000	90	100	47
8	43	Asia	Granulosa Cell Ca ovary	–	+	+	+	–	67	933	90	14	35
9	39	No Afr	Adeno Ca ovary	–	+	–	–	–	78	666	95	9.5	63
10	49	Eur	Adeno Ca ovary	–	+	–	–	–	70	600	84	0	21
11	67	Eur	Mucinous Cystadeno Ca ovary	+	+	–	–	–	79	600	95	15	49
12	44	Asia	Ca of Cx	+	–	+	–	–	64	800	87	8	16
13	77	Eur	Ca of endometrium	–	+	+	–	–	69	866	76	40	27
14	41	Asia	Granulosa cell Cx ovary	–	+	+	–	–	76	666	71	29	44
15	73	Asia	Adeno Ca Cx	–	+	–	–	–	65	667	92	39	55
Normal													
1	M/48	US							65	1,267	94	30	38
2	M/66	Asia							48	1,267	86	1	23
3	F/62	USSR							68	933	78	17	33
4	M/53	Eur							70	866	85	3	47
5	M/52	Asia							68	933	86	19	40
6	F/58	Eur							78	800	81	57	29
7	F/40	Asia							73	866	92	9	25
8	M/43	No Afr							66	1,066	93	–2	54
9	M/33	Eur							71	933	88	6	36
10	F/62	USSR							65	933	87	–4	49
11	F/55	Asia							68	1,066	87	–13	22

12	M/44	Asia		73	933	93	12	32
13	F/37	Eur		78	866	86	16	48
14	M/46	No Afr		75	1,133	91	0	30
15	M/42	Eur		71	933	93	−15	11
16	F/52	Eur		73	866	91	0	83
17	F/22	Eur		80	667	60	0	62
18	F/40	Eur		68	1,333	97	−11.4	21
19	F/55	Eur		65	667	95	20	4
20	M/43	Asia		70	1,000	−6	−18	87
21	M/50	Asia		74	800	77	3	59
Ovarian carcinoma, diagnosed and tested before treatment								
1	70	Asia	Serous Adeno Ca	71	1,200	71	−10	
2	52	No Afr	Endometroid Adeno Ca	65	1,100	84	25	
3	22	USSR	Serous Adeno Ca	63	1,115	74	−2	
4	55	USSR	Serous Adeno Ca	70	584	75	−7	
5	77	Eur	Serous Adeno Ca	68	396	90	25	
6	61	No Afr	Serous Adeno Ca	72	904	95	29	
7	57	USSR	Mucinous Adeno Ca	68	1,000	45	13	
8	60	USSR	Serous Adeno Ca	70	1,100	90	23	
9	40	Asia	Serous Adeno Ca	72	720	95	5	
10	42	Asia	Endometroid Adeno Ca	75	590	85	11	
11	42	Asia	Granulosa cell	69	860	95	10	

Ca = carcinoma; Eur = Europe; No Afr = North Africa; Inf Duct = infraductal; Bil Comedo = bilateral comedo carcinoma; Bil Adeno = bilateral adenocarcinoma; In Op = inoperable; Sq = squamous; Dif = differentiated; Pros = prostate; Kid = kidney; Cx = cervix; Cystadeno Ca = cystadenocarcinoma; Conc = concentration; ChE Sp Act = cholinesterase specific activity.

Description of clinical treatment includes surgery (Surg), irradiation (Irrad), chemotherapy (Chem) and hormone administration (Horm). Protein concentration was determined on two serum dilutions [41]. The specific activity of ChE was measured by [^{3}H]-ACh hydrolysis [166]. Inhibition values were derived from complete (seven-point) inhibition curves in each case. In cases where negative inhibition values are noted, the activity measured was higher in the presence of inhibitor. This is probably due to the action of the inhibitor on other, perhaps, competitive, serum proteins (reproduced from Zakut et al. [358], with permission).

from controls, with high sensitivity towards iso-OMPA and considerably lower sensitivity to BW284C51; (2) curves displaying high sensitivity to BW284C51 and normal inhibition by iso-OMPA, and (3) curves with the same level of sensitivity towards both inhibitors [358].

In some of the samples examined, low concentrations of particular inhibitors caused an increase in the rate of ACh hydrolysis. This may be due to the blocking action of these inhibitors on the competitive or inhibitory activity of other serum proteins.

It is noteworthy that about 75% of the samples examined were drawn postsurgery from patients under irradiation therapy, chemotherapy or hormone treatment. In this respect the samples varied from previously analyzed serum samples (i.e. those of the H family, see above) and from DNA samples of patients with platelet disorders. Metastases were diagnosed in a few patients only, and the age group examined ranged between 18 and 90 years. Specific values from the above-described inhibition curves (the total ChE-specific activity and the inhibitions caused by 50 ng/ml succinylcholine, 1×10^{-5} *M* BW284C51 and 1×10^{-4} *M* iso-OMPA) were selected as representative data. In most of these measurements, the total ChE activity was lower and the sensitivity to BW284C51 higher in tumor samples as compared with controls. In contrast, there were no differences in the total protein concentration or in the sensitivity to iso-OMPA or succinylcholine. The enhanced sensitivity to BW284C51 was not significantly correlated to age, sex, ethnic origin or mode of treatment, and did not differ between patients suffering from distinct types of carcinomas (table 3). Because of insufficient histological information, it remained unclear whether the modified properties are related to the state of differentiation of particular carcinomas.

Summary of the biochemical observations revealed that the average specific activity of ChE was significantly lower by about 25% ($p < 0.001$) in the serum of tumor patients, in agreement with previous findings of others [reviewed in 347]. This phenomenon is generally attributed to decreased functioning of the liver in patients under antitumor therapy. The sensitivity to iso-OMPA was generally similar in tumor serum samples to that measured in control samples. In contrast, 1×10^{-5} *M* BW284C51 was sufficient to cause a $29 \pm 6\%$ decrease in ChE activity in the serum samples from patients but only $6 \pm 16\%$ in controls ($p < 0.0005$). These included two samples in which BW284C51 inhibition was exceptionally high [358]. Succinylcholine, in turn, showed relatively low variability in the extent of inhibition in tumor serum samples. However, there was no significant difference between the inhibition observed in diseased and healthy sera, and 50 ng/ml of this analog was essentially sufficient to cause substantial block in all serum samples (table 4).

Table 4. Properties of cholinesterases in serum of carcinoma patients

No.	Type of carcinoma under treatment	Number of samples/ patients	Age distribution (mean)	Protein conc. mg/ml	ChE sp. act. $\left(\frac{\text{pmol}/\mu\text{g P}}{\text{h}}\right)$	$+1.10^{-4}$ *M* iso-OMPA	$+1.10^{-5}$ *M* BW284C51	+50 ng/ml Suc Ch
1	Breast	28	30–71(54.9)	72.3 ± 8.3	667 ± 333	83.9 ± 16.1	24.1 ± 21.4	35.3 ± 43.7
2	Lung	7	43–82(61.0)	72.3 ± 6.5	667 ± 200	88.4 ± 6.9	39.6 ± 19.8	30.6 ± 14.8
3	Digestive tract	17	42–86(67.2)	65.8 ± 10	800 ± 266	88.5 ± 9.5	31.8 ± 22.4	25.5 ± 34.3
4	Urinary tract	10	70–86(73.6)	71 ± 5.5	800 ± 266	79.3 ± 11.7	27.1 ± 21.3	26.9 ± 20.2
5	Gynecological	15	39–73(57.0)	71.5 ± 5.5	733 ± 133	85.8 ± 9.3	23.6 ± 27.0	39.0 ± 18.2
6	Average of all treated tumor types	–	30–86(61.0)	70.7 ± 2.2	733 ± 59	85.2 ± 3.4	29.2 ± 6.0	31.5 ± 5.0
7	Untreated ovarian carcinomas	11	22–70(52.5)	70.2 ± 1.8	799 ± 58	73 ± 10.1	9.3 ± 12.5	n.d.
8	Control	21	22–66(47.8)	69.8 ± 6.5	960 ± 175	83.0 ± 19.0	6.2 ± 16.6	39.2 ± 20.6
	Total	109						

Average values and standard deviations were calculated for each of the test groups separately and for all of the tumor samples together. Statistical significance (p) was measured by the Student's t test and is noted for the decrease in ChE specific activity and for the enhanced sensitivity to **BW**. ChE sp. act. = cholinesterase specific activity; conc = concentration; **BW**284C51 = 1,5-bis(4-allyldimethylammonium-phenyl)pentan-3-one dibromide (reproduced from Zakut et al. [358], with permission).

Analysis of the distribution of modified ChE properties in various tumor samples indicated that the enhanced sensitivity to BW284C51 was not related to the decrease in total ChE activity or to the stable susceptibility to iso-OMPA (fig. 16). This implied that the increased sensitivity to BW284C51 could reflect the appearance of another form of ChE in the serum samples from cancer patients. To further test this possibility, several serum samples were subjected to sucrose gradient centrifugation and determination of the sedimentation profile of ChE forms in these samples in the absence of presence of BW284C51 and iso-OMPA. The major ChE form in all of these gradients displayed a sedimentation value of ca. 12S, as expected from soluble BuChE tetramers and in agreement with previous investigators [eg. see 274]. The 12S form was the principal one in control serum samples. In contrast, we detected an additional minor but reproducible peak of activity sedimenting as ca. 6–7S in gradient profiles of

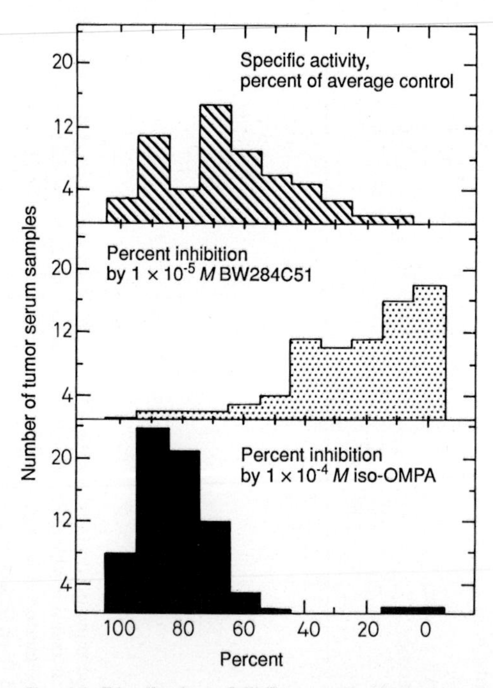

Fig. 16. Distribution of ChE properties in tumor serum samples. Values follow the data presented in tables 3 and 4. The figure presents percent of tumor serum samples (of a total of 77) which displayed certain ChE activities calculated as percent of average control activities (top), as well as percent of tumor serum samples which exhibited certain inhibition by 1×10^{-5} *M* BW284C51 (middle) or 1×10^{-4} *M* iso-OMPA (bottom) [reproduced from 358, with permission].

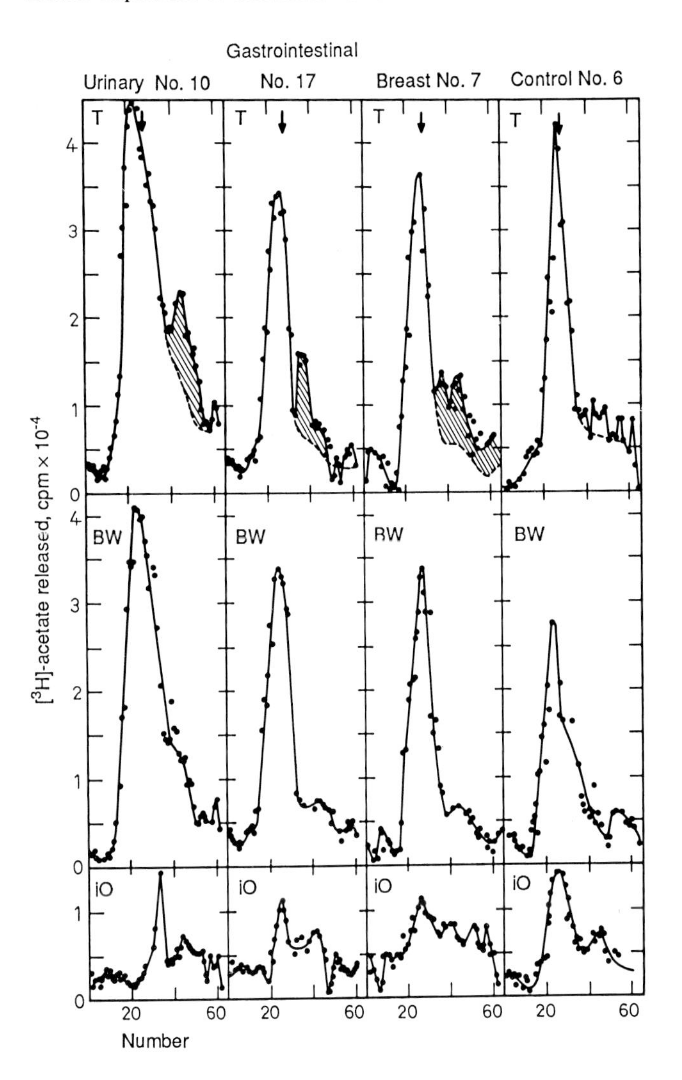

Fig. 17. Sucrose gradient profiles showing molecular forms of ChEs in diseased and control serum samples. Serum samples were centrifuged on 5–20% sucrose gradients, and analyzed for ChE activity in the presence and absence of appropriate inhibitors. The type of tumor and number of sample are noted for each profile. T = total ChE activity; BW = activity in the presence of 1×10^{-5} *M* BW284C51; iO = activity in the presence of 1×10^{-4} *M* iso-OMPA. The tumor-characteristic peak of ChE activity susceptible to both inhibitors is marked by dotted lines on the top gradient profiles [reproduced from 358, with permission].

serum samples from patients suffering from various tumor types. This tumor-characteristic ChE form had the same sedimentation coefficient as the BuChE dimers produced from recombinant BuChE mRNA in micro-injected *Xenopus* oocytes. However, it differed from the recombinant enzyme in that it was quantitatively blocked by both BW284C51 and iso-OMPA, which implied that it was neither AChE nor BuChE, but another form of ChE with combined properties of both (fig. 17). This form of ChE could be responsible for the considerable enhancement in sensitivity to BW284C51 in serum ChEs which also displayed normal sensitivity to iso-OMPA.

III. Basic Research and Clinical Implications

1. Biochemical Implications to Sequence Similarities within the Cholinesterase Family

Cholinesterases have a catalytic mechanism similar to that known for the serine proteases [reviewed in 260]. In line with the widely accepted notions that functional similarity reflects common ancestral genes and that multigene families have developed by gene duplication and subsequent divergence during evolution [249], Augustinsson [20] suggested that ChEs are phylogenetically related to the large family of serine proteases and may be defined as members in the multigene family of serine hydrolases. The 3 out of 8 match in the consensus sequences of the catalytic sites of carboxylesterases and serine proteases, including the invariant serine residue, further suggested a common origin to these two families [244]. Recent molecular cloning and DNA-sequencing studies confirmed the phylogenetic relationships within the gene families of serine proteases [278] and of carboxylesterases [154], but left the question of their interrelationships open.

Profile analysis of the human AChE amino acid sequence showed no similarity to any of the 4,500 protein sequences in the European Molecular Biology Laboratory (EMBL) protein data base, with the exceptions of *Torpedo* and *Drosophila* AChEs, human BuChE and bovine Tg [232]. Human AChE did not specifically show any resemblance of serine protease sequences. In view of the sequence information deduced from cloned DNAs for four different ChEs, and based on the above discussed arguments, it now appears that human AChE joins other ChE species to form a limited minigene family that belongs to the larger family of carboxylesterases type B but appears to be distinct from the other serine proteases. Our analysis therefore extends and supports the recent conclusion of Richmond et al. [241] in suggesting that ChEs cannot be included in a serine hydrolase multigene family.

Within the ChE family, the high amino acid sequence similarities between human AChE and BuChE imply that variations and conservations in the primary amino acid sequence of ChEs may be implicated with distinct differences in the substrate specificity and sensitivity to selective inhibitors that were observed for particular types of ChEs. Detailed analysis of these sequences by site-directed mutagenesis and expression of the modified genes in heterologous systems may therefore reveal the key residues in the charge-relay system of ChEs and lead to the development of improved therapeutic drugs against OP intoxication [237]. Interestingly, the homology between ChEs is considerably higher at the amino acid level than it is at the level of nucleotides. Furthermore, the A,T-rich composition of BuChE cDNA as opposed to the G,C-rich composition of AChE cDNA indicates a different structural folding pattern [230] as well as a more ancient origin for the CHE gene, in complete accordance with the ontogenetic coordination in the developmental expression profile of the two cholinesterases.

Screening of several cDNA libraries from various tissue origins with OPSYN-active site probes [315] resulted in the isolation of identical cDNA clones, all coding for serum BuChE. These were apparently transcribed from the same gene and were processed similarly both in fetal and adult tissues. The relatively low number of AChE cDNA clones in these libraries was later explained by the densely packed secondary structure of AChE cDNA [Soreq et al., in preparation]. Also, these findings indicate that the human ChE gene, encoding serum BuChE, is most probably transcribed into a single transcription product in all tissues, with the exception of alternative termination [Gnatt et al., unpubl.]. We cannot as yet exclude the possibility that alternative splicing is involved in the regulation of this gene in humans, similarly to its use in AChE production in *Torpedo* [300, 301]. However, our present findings imply that alternative splicing is not responsible for the heterogeneity observed in liver and brain BuChE forms. In the absence of indications for the involvement of transcriptional control, posttranscriptional mechanisms can be pursued as an origin for BuChE heterogeneity. This assumption was supported by the oocyte microinjection experiments with recombinant synthetic BuChE mRNA.

2. Posttranscriptional Regulation of Cholinesterase Heterogeneity

When injected into *Xenopus* oocytes, synthetic mRNA transcribed off a cDNA-encoding human serum BuChE induced the production of a protein displaying the substrate specificity and sensitivity to selective inhibitors characteristic of native BuChE, and which clearly distinguish it from AChE. These results indicate that the ligand-binding specificities of BuChE reflect a property which is inherent to the primary amino acid sequence of the molecule. The level of mRNA expression achieved in the oocytes using pure synthetic message was relatively low, when compared with the 10^5-fold enrichment of BuChE mRNA over that observed in tissue extracts [264, 312]. This suggested that the injected BuChE mRNA is encountered with competing endogenous oocyte mRNAs for the limited number of rough endoplasmic polysomes in the oocyte [313]. Alternatively, or in addition, it could reflect limited distribution of the injected RNA within the oocyte.

The 10-fold increased IC_{50} toward iso-OMPA observed for the membrane-associated enzyme demonstrated that interaction with detergent at the extraction step alters its hydrophobic character, which reduced its affinity for this OP ligand. Similar alterations occurred when detergent was added to the soluble oocyte fractions [319]. However, the K_m for BuSCh did not appear to be altered by detergent association. The observation that the single mRNA template employed directed the synthesis of catalytically active BuChE displaying the native ligand-binding affinities could exclude post-translational modifications in defining the 3-dimensional structure of the active site and suggested that the variable BuChE phenotypes, like the 'atypical' or the fluoride-insensitive serum enzymes [348], reflect mutations altering the amino acid sequence of these enzyme forms. However, it should be noted that such changes should not necessarily occur in the active site region, as indeed is the case for the 'atypical' BuChE enzyme [227], reported as caused by a single alteration of Asp 70 into Gly. More recently, several point mutations were found in our laboratory and were shown to induce significant differences in the biochemical properties of recombinant BuChE forms produced in microinjected oocytes [Neville et al., unpubl.].

Taking the average membrane-associated activity as 8.1 μmol BuSCh hydrolyzed/h/oocyte implies 1.2×10^{15} molecules BuSCh hydrolyzed/s/oocyte. Assuming a turnover rate of 1×10^4 molecules/active site/s [63] implies 1.2×10^{11} catalytic sites/oocyte in the membrane-associated fraction, or 6×10^{10} dimeric BuChE molecules/oocyte. In an oocyte injected with synthetic BuChE mRNA alone, approximately 2,000 μm^2, or 0.01% of the oocyte surface, was occupied by high-density immunoreactive ChE molecules [105]. Outside these areas the density of BuChE molecules on the cell surface is therefore suggested to be below the 1×10^3 molecules/μm^2 threshold required for detection using immunohistochemical techniques [290]. We therefore concluded that the nonfluorescent surface area contained undetected BuChE molecules evenly distributed at a concentration 10-fold lower than that found in the immunoreactive loci. Assuming that the total external surface area of the oocyte actually includes the theca and follicle cell layers as well as the enlargement contributed by the micro- and macrovilli of the oocyte and follicle membranes [107, 349] yields a 100-fold enlargement of the total surface area sequestering BuChE and allows us to consider a factor of 10^5 in the relative area occupied by undetected as opposed to immunostained ChE. Therefore, the molecular density of ChE in a cluster or patch can be estimated to be: $6 \times 10^{10} = 2 \times 10^3(X) + (0.1X)(2{,}000)(10^5)$ or, $(X) = 3 \times 10^3$ molecules/μm^2.

It is interesting to note that both the nicotinic ACh receptor [290, 326] and AChE in neuromuscular junctions and along neuronal dendrites [156, 285] were estimated to be aggregated at molecular densities within the same order of magnitude. This observation might indicate that the organization of membrane-bound molecules within the extracellular surface reflects a precisely regulated physiological property of the involved subcellular structures which is conserved through evolution.

Microinjected alone, synthetic BuChE mRNA induced the formation of primarily dimeric ChE. This first level of oligomeric assembly may therefore be spontaneous, or may require a catalytic mechanism that is already available in the oocyte. It should be noted in this respect that the detection of ChE activities in *Xenopus* oocytes [313] and high levels of BuChE mRNA in human oocytes [316] indicate that ChE represents a natural endogenous oocyte protein. The observation that tissue-extracted mRNAs induced higher levels of multisubunit assembly – possibly including the incorporation of noncatalytic subunits – indicates that additional protein species, not available in the oocyte, are required to direct the biosynthesis of more complex molecular forms. Given the high degree of tissue-specific polymorphism of ChE molecular forms, it is not surprising to find these additional factors expressed in a tissue-specific manner. Nonetheless, the nature and number of these factors remains to be elucidated.

Supplementation with tissue mRNAs increased the number and intensity of patches and clusters indicating the induction, by tissue-specific factors, of enhanced BuChE aggregation at the external surface of the oocyte. This property of enhanced aggregation seems to be correlated to the appearance of molecular forms containing 4 or more catalytic subunits, and may reflect the provision of tissue-specific membrane-anchoring elements. Such elements were found to be required for the aggregation of both the nicotinic ACh receptor and AChE at the neuromuscular junction [12, 343, 344]. It would be intriguing to reveal whether the requirements for BuChE are parallel or whether the two nascent proteins compete in vivo for a limited pool of noncatalytic elements.

In contrast with other genes (e.g. EGF) [53] it appears that the altered mode of BuChE expression in oocytes coinjected with tissue mRNAs did not result from altered posttranslational processing of nascent polypeptides but involved the interaction with other cell-surface molecules. Although both brain and muscle mRNAs induced both patches and clusters on the oocyte surface, the relative distribution of each type of formation varied between oocytes coinjected with mRNA from the two tissue types in a manner consistent with the organization of ChE in the native tissues [235, 343, 344]. Furthermore, the relative staining intensity of clusters and patches obtained with muscle mRNA qualitatively exceeded that obtained with brain mRNA. Together, these observations imply a qualitative and/or quantitative difference between ChE-related mRNAs in different tissues and that these mRNA pools are capable of modulating tisssue-specific use of a single BuChE mRNA species.

Several classes of membrane-anchoring elements have been considered in conjunction with AChE. Considerable evidence implicates covalently linked glycolipid, most prominently phosphatidylinositol, as a principal mechanism for the anchorage of globular dimeric AChEs in plasma membranes [277, 304]. Noncatalytic subunits and proteoglycans have been implicated in the attachment of AChE to membranes in the brain [163] and to the basal lamina at the neuromuscular junction [42, 101, 102], where aggregates of ACh receptors were also shown to be associated with plaques of heparan sulfate proteoglycan [10]. We do not yet know which of these mechanisms, if any, is operative for BuChE in our system. However, the observation that alternative mRNA processing is involved in the membrane attachment of hydrophobic AChE in *Torpedo* [135, 136, 300, 301] does not appear applicable here since a single mature BuChE mRNA induced a complete array of molecular forms in the oocytes including soluble and nonsoluble pools.

Taken together, these findings demonstrate that both muscle and brain express mRNAs encoding peptides which: (1) are required for the biosynthesis of BuChE molecular forms consisting of multiple dimeric units; (2) may specify the incorporation of noncatalytic subunits; (3) direct the formation of a tissue-specific array of BuChE molecular forms, and (4) direct a tissue-specific organizational pattern of BuChE at the external surface of the injected oocytes.

The conspicuous intracellular accumulation of active enzyme induced by tunicamycin indicates that posttranslational glycosylation is not required for the enzyme's activity but plays an essential role in the transport of catalytically active BuChE to the external surface. The unequal animal-vegetal pole distribution of induced BuChE in the oocytes indicates an active, polar asymmetry which has been described previously for morphological characteristics such as the yolk platelets or cytoskeletal elements [82, 107, 257, 349] and also for a maternal mRNA localized in the vegetal hemisphere [346]. Our findings suggested that the polar buildup of the oocyte cytoskeletal elements takes an active part in directing newly synthesized proteins to their ultimate site of association at predefined extracellular positions in a manner similar to other polarized cell types [128] and that the animal pole of the *Xenopus* oocyte is preferentially designated for depositing nascent surface-associated proteins. Parallel mechanisms in muscle fibers may explain the site-directed accumulation of ChEs in neuromuscular junctions, and microinjected *Xenopus* oocytes may hence provide a most appropriate model system for studying these subcellular transport mechanisms.

3. Immunochemical Implications to the Similarity and Heterogeneity between Cholinesterase Forms in Various Tissues

BuChE represents one member of a family of enzymes. To better understand the interrelationships between members in this family, we examined its immunochemical similarity to the other principal member of this family, AChE, as well as to the sequence-homologous protein Tg, by immunoprecipitation and immunocytochemical approaches in several model systems.

For this purpose, peptides included in the 200 N-terminal amino acids of human BuChE were used as an antigen. This part of the enzyme may be distinguished both by its particularly low immunogenicity, as revealed by the computer prediction according to Chou and Fasman [68] and Devereux et al. [91], and by its considerable sequence homology to other ChEs [315]. The cDNA-produced BuChE peptides, having been produced in bacteria, were not glycosylated. Furthermore, these polypeptides include only part of the cysteine residues forming the S–S bonds in the complete BuChE molecule, and their tertiary structure is, most probably, different from that of the native protein.

Polyclonal rabbit antibodies were elicited against these naked, processed peptides, with the following assumptions in mind; (a) The anti-recombinant BuChE antibodies might have novel properties, not detected previously, since conventional antisera, induced against the mature, fully folded enzyme, would not include antibodies against the low immunogenicity N-terminal part of BuChE as major species. (b) Antibodies elicited against this particular part of the enzyme would tend to have a low titer to the antigen because of its low immunogenicity, and might display low affinity because of the complete lack of processing of these peptides in bacteria. (c) Such antibodies would interact with particular forms of the native, nondenatured enzyme only if the epitopes within the N-terminal polypeptide would be at the external surface of the antigen molecules and accessible for immunoreaction. Consequently, these domain-specific antibodies could be useful for the detection of both sequence homologies and structural differences between various molecular forms of ChEs, in spite of their expected low affinity binding and low titer. The heterologous patterns of immunoreactivity obtained in this study confirm the validity of this assumption.

When tested by immunoblot analysis, the recombinant BuChE peptides produced in bacteria interacted efficiently with antibodies reported to be selective for AChE [222, 309]. Vice versa, the antirecombinant BuChE antibodies interacted efficiently with the blotted fully denatured forms of both BuChE and AChE from blood. In contrast, their ability to precipitate the native or even denatured enzyme from solution was rather poor and with higher specificity towards BuChE, in agreement with the above noted expectations. Even in the presence of a second antibody, which binds all of the IgG molecules present in the immunoreaction mixture, an equivalence zone of 1:80 dilution was determined [104]. Assuming that the concentration of the antibody in the equivalence zone is close to the K_d, this dilution implies that the concentration of active antibodies in raw antiserum is about 1×10^{-7} *M*. These results could reflect the low immunogenicity of the N-terminal part of the BuChE molecule, as predicted by computer analysis.

About 10–15% of the total activity could be precipitated under these dilution conditions. Because the rabbit immunization was performed with a denatured ChE, produced in bacteria from a cDNA clone, the specificity of the immunoreaction was also examined in solution with the denatured purified enzyme, previously labeled by ^{3}H-DFP. The antibody titer remained equal to that observed with the native enzyme, confirming that there is a very low concentration of the specific antibody.

Sucrose gradient centrifugation revealed that the antibodies interacted specifically with BuChE tetramers from serum and BuChE dimers from fetal muscle. They also bound to dimers and tetramers of AChE from fetal muscle. However, there was no significant binding to the AChE dimers from the erythrocyte membrane or to tetramers of BuChE from fetal muscle. The common element to sedimentation peaks which include ChE molecular forms that interact with the antibodies is that they all contain hydrophilic forms. Furthermore, the extent of precipitation in muscle extracts could correspond to the fraction of hydrophilic forms within these particular peaks. These results therefore suggested that our polyclonal antibodies were able to recognize epitope(s) that are common to or exposed for immunoreaction, only in hydrophilic ChE molecular forms.

The antibody-induced alteration of the sedimentation coefficient observed in our gradient analyses (2S) was rather low compared to the values given in the literature (about 3S) [119, 222, 269, 309, 310]. This finding as well was in accordance with the low titer of the antibody for the serum

BuChE, as only 15% of the molecules were shifted. It is impossible to detect two peaks of reacted and nonreacted enzyme molecules on a gradient where the standard deviation is of the same order of magnitude as the shift. In fetal muscle extracts, the dimer peak alone represents 45% of the total activity; thus the proportion of antibody-bound enzyme was higher in this form, which explains its clearer shift.

In addition to their high efficiency in immunoblots, antirecombinant ChE antibodies were also used for labeling the bound enzyme in muscle fibers. Several lines of evidence suggest that the regions labeled by these antibodies in crushed muscle fibers are indeed the endplate ones: (a) In all of the fibers analyzed, we never found more than one labeled region per fiber, which is in accordance with the constant monofocal innervation of human skeletal muscle. (b) The morphology of the binding region, a very simple 'plaque', suggests a single gutter structure without secondary foldings, which corresponds to the premature neuromuscular junctions that one expects to observe at this developmental stage. (c) In all cases, labeling coincided with intensive cytochemical staining for ChE activity, as expected for endplate regions. Based on this evidence, the in situ labeling of slightly fixed crushed fetal muscle fibers using these antibodies suggests that the N-terminal part of human BuChE or a peptide highly homologous to it is present in an immunoreactively exposed form in the fetal junction. Although we cannot conclude as yet whether the in situ binding occurs with AChE or BuChE, both of which are highly concentrated in endplate regions, it is interesting to note the correspondence between immunoreactivity and ChE activity at the neuromuscular junction. This most probably implies that ChEs in neuromuscular junctions are produced in loco and are not transported in the form of nascent-inactive peptides from other sites along the muscle fibers.

In conclusion, the experiments performed with antibodies elicited towards the N-terminal part of BuChE as expressed from cloned cDNA in bacteria suggested the existence of extensive sequence homology in this part of human ChEs between various forms of AChE and BuChE in different tissues. Previous reports of the lack of cross-reactivity between antibodies to AChE and BuChE could hence be explained by the low immunogenicity in this part of the molecule or be due to structural differences between various molecular forms of the enzyme within and between tissues. It should be noted in this respect that the complete amino acid sequence of human AChE has recently been determined in our laboratory [Soreq et al., in preparation], and that this conclusion has been fully confirmed by the data translated from the cDNA coding for AChE.

Our findings also point out that similarities in substrate specificity and molecular form composition (such as those between BuChE tetramers in blood and BuChE tetramers in fetal muscle) do not exclude the possibility of structural differences between such homologous forms. This is in agreement with the evidence on variable mRNAs affecting the molecular form heterogeneity of BuChE in a tissue-specific manner [297]. Posttranslational differences would explain the heterologous immunochemical properties of highly similar BuChEs and emphasize the importance of such processing to the final buildup of the polymorphic ChEs in different tissues and body fluids.

4. Autoimmune Anticholinesterase Antibodies May Be Implicated in Graves' Ophthalmopathy and Muscle Disorders

Similarly to other proteins, ChEs may induce autoimmune antibodies under pathological situations. Because of the fundamental importance of ChEs to muscle functioning, such antibodies may cause neuromuscular disorders. This may explain why a patient with autoimmune anti-ChE antibodies [206] presented with clinical and laboratory findings that were quite distinctive from those observed in other recognized diseases characterized by chronic muscle weakness and a fluctuating course [1]. The findings of low levels of AChE, combined with some clinical features resembling those of OP poisoning [181], led us to presume a role for AChE inhibitors in the etiology of this disorder. An extensive search failed to reveal the presence of AChE-blocking agents, and AChE antibodies were looked for instead. Minute quantities of the patient's serum inhibited and precipitated the activity of AChE from various human tissues, in a fashion similar to that of rabbit antibodies against rat brain AChE, as was reported by Marsh et al. [222]. The membrane-associated fraction of muscle AChE was preferentially inhibited. The inhibition of the enzyme resulted from Ig and not from protease degradation or binding to a chemical. This was indicated by the abolishment of the inhibition upon preabsorption of the patient's serum with goat anti-human Fab and by the capacity of the patient's serum to precipitate the enzyme. Altogether these findings indicated that the patient's serum contained anti-ChE antibodies.

It is reasonable to assume that these anti-AChE antibodies led to the low levels of BuChE in the patient's serum and AChE in his red blood cells. Other conditions that may give a low level of AChE were all excluded. These include chemical inhibition, hereditary defect, paroxysmal nocturnal hemoglobinuria, pernicious anemia, and thalassemia. Of those, the only relevant one that gives very depressed AChE levels and affects both serum and red blood cells is chemical inhibition, and this was thoroughly studied and ruled out.

It is also reasonable to assume that the AChE in the neuromuscular junction served as another target for the patient anti-AChE antibodies. In experimental models it was found that despite the large variability of enzyme forms, antibodies raised against one type of AChE cross-react with all the other types [45, 104, 119, 222]. This also implies to the above described case, where the patient's serum could inhibit AChE derived from six different tissues studied. The patient's serum interacted preferentially with membrane-associated AChE from human muscle, thus making the possibility of such an antigen-antibody response in the patient's muscle tissue very likely. The ability of antibodies to penetrate the endplate in order to interact with various constituents of the neuromuscular junction is a question already resolved. This occurs in myasthenia gravis [205], and in an experimental model of antibodies to choline acetyltransferase [62]. Our patient exhibited high sensitivity to edrophonium, resulting in total loss of skeletal muscle control. This suggests that most of the AChE in the neuromuscular junction was already inactive, presumably due to the antibody effect.

During the course of the disease, signs of neuropathy, myasthenic responses, and muscular atrophy rapidly developed, as was observed in the biopsy results and in the electrodiagnostic studies. All these developments were explainable by antibody effect alone. The immediate and the primary result of AChE inhibition in the neuromuscular junction is a disturbed neuromuscular transmission. This is due to the accumulation of ACh in the synaptic cleft leading to persistent depolarization and subsequent desensitization of the postsynaptic membrane [334]. In addition, inhibitors of AChE, especially if chronically applied, may cause muscle and nerve injury, either by direct unidentified mechanisms [118, 193, 335, 345] or by secondary adaptation responses as was demonstrated in other neuromuscular disorders [95, 120, 238]. Finally, antibody directed to an enzyme in the neuromuscular junction can cause extensive damage in the neuromuscular junction. This was shown in the experimental disease produced by antibodies to choline acetyltransferase [62]. The clinical course and the laboratory findings in this case were therefore in agreement with what might be expected from anti-AChE antibodies.

The causal relationship between anti-AChE antibodies and the patient's disease may be argued. One may suggest, for instance, a primary insult to the muscle tissue and its components including AChE and a secondary response of 'accompanying' antibody formation. We feel that the combination of very low to zero levels of AChE, the presence of antibodies to AChE, and a clinical course of progressive, chronic fluctuating muscle weakness identify a separate entity in muscle diseases. The role of autoantibodies in many connective tissue diseases (anti-DNA antibodies in lupus erythematosus, rheumatoid factor in rheumatoid arthritis, etc.), although considered important, is not completely clear. How serology correlates with disease severity and activity is not known. In this respect the disorder discovered in our patient is not an exception. Until the role of the anti-AChE antibodies is made clear, they may be considered a diagnostic finding in this disease.

Phillips et al. [261] studied a case of a patient with muscle weakness diagnosed as suffering from myasthenia gravis. In that case, antibodies to AChE were found in the neuromuscular junction of the patient's muscles and in the patient's serum. The antibodies in the serum were shown to react with human erythrocyte AChE, and with the patient's neuromuscular junction. A role for the antibodies in the patient's disease was postulated. Unfortunately the authors did not provide a detailed case history and there was no mention of serum or red blood cell AChE levels. Their patient resembled ours in her susceptibility to anti-ChE medications, in the myasthenic features of the EMG studies, and in the light microscopy studies. It is possible that both cases had the same disease and Phillip's patient was diagnosed differently because of the myasthenic feature of this disorder in the EMG records. Alternatively, Phillip's patient could have been myasthenic as suggested. In that case, to the best of our knowledge, our patient will be the first described with this unique disease.

How frequent are autoimmune antibodies against ChEs, and in which diseases would one expect them to occur? The interaction of polyclonal antibodies raised against the N′-terminal part of recombinant BuChE with Tg clearly demonstrated that cross-reactivity exists between antibodies to these two proteins in their reduced and denatured state. Furthermore, we have demonstrated that Igs from patients with Graves' ophthalmopathy will bind to ChE in a variety of conditions. Anti-ChE antibodies did not always correlate with anti-Tg activity in the protein blot experiments, contrary to preliminary results reported using an ELISA in which there was a strong correlation between the two [212, 213, 331]. However, it is important to note in this respect that the antibodies used in the more recent study were raised against a naked, nonglycosylated part of the human BuChE protein whereas those tested by ELISA were elicited against the mature, fully glycosylated AChE from *Torpedo*. It is notable that binding to Tg was not always detected, even amongst patients with higher titers of anti-Tg antibodies as measured by ELISA. It may be that an epitope, recognized by these antibodies, is either prominent in the clone-produced part of the human enzyme or modified in the reducing conditions of the SDS-PAGE.

Not all patients with Graves' ophthalmopathy have detectable levels of anti-Tg antibodies, and yet two antisera in this study that were anti-Tg negative by ELISA bound to ChE in a protein blot. These sera may contain antibodies with a low affinity for Tg but of higher affinity for BuChE. Consequently, they bind to BuChE despite being beyond the limits of detection using Tg.

Immunolocalization studies demonstrated that the binding to ChE observed in patients with Graves' ophthalmopathy was not merely an in vitro phenomenon. Previous attempts to demonstrate antibodies to eye muscle using a variety of methods, both in vitro [299] and in situ, have been largely unsuccessful with the exception of Mengitsu et al. [231] who have shown diffuse cytoplasmic staining using Graves' ophthalmopathy sera by immunofluorescence.

We have focused on formaldehyde-fixed endplate regions of muscle since the antiserum used was shown, as detailed above, to interact with blotted denatured ChE and because ChE is present at higher concentrations in these areas although some forms of ChE are also found extrajunctionally. In fetal muscle, the basal lamina is not fully mature and so ChE is more accessible for antibody binding. Clearly the experiment would be more conclusive using extraocular muscle, although this poses some technical difficulty because of the limited quantity of such tissue available [106].

If anti-ChE antibodies are implicated with ocular muscle pathology, this indicates that ChEs are important for the normal development and/or functioning of various cell types. Another aspect of this conclusion is reflected in the cases of the H family, and the leukemic DNAs, where amplification of the CHE and the ACHE genes, was observed (see Sections III.6 and III.7).

5. Expression of Cholinesterase Genes in Haploid Genome Suggests an Involvement in Germline Cells Development and/or Functioning

The continuously high expression of BuChE mRNA throughout oocyte development suggests that the enzyme may be required for oocyte growth and maturation processes. ACh promotes progesterone-induced maturation of *Xenopus* oocytes [84]. Furthermore, inositol 1,4,5-triphosphate, which mimics muscarinic responses in *Xenopus* oocytes [253], triggers the progesterone-induced activation of amphibian and starfish oocytes. AChE in *Xenopus* oocytes is seasonally regulated inversely to the reproductive cycle [313], similarly to the muscarinic receptors in these oocytes [83]. Damage to the entire cascade of the ChE gene family in nematodes is lethal to the animals [170]. Altogether, this may indicate an involvement of cholinergic responses in the process leading to meiotic maturation of oocytes in a selected group of antral follicles. Alternatively, or in addition, the enhanced transcription of ChE genes in oocytes may reflect the accumulation of BuChE mRNA for later use during the postfertilization processes, similar to the accumulation of excess histone mRNAs in developing sea urchin oocytes [352]. This is in good agreement with findings demonstrating transiently enhanced BuChE activity early in chick embryogenesis [195, 197–199]. Finally, it is possible that AChE is involved in sperm-egg interaction or postfertilization mechanisms, as suggested from the cholinergic induction of polyspermy in sea urchin oocytes [272]. Elicitation of transgenic mice carrying the transgene BuChE cDNA sequence will assist in the search for the specific role(s) of BuChE in fertilization processes.

The pronounced synthesis of BuChE transcripts in oocytes suggests that the CHE genes in humans are particularly good candidates for the formation and reinsertion of inheritably amplified genes. Also it would be important to determine to what extent the hybrids performed in situ are perfect or whether unpaired non-Watson-Crick base pairs are involved [117]. Further characterization of these genes will be required to correlate between specific copies of CHE genes and the particular BuChE mRNA transcripts produced in the developing oocytes. For example, it will be interesting to find out whether specific alleles of the CHE gene (such as the 'atypical' or the 'silent' ones [348] produce enzyme forms that are more susceptible than others to ecological exposure to OP inhibitors. If so, one wonders whether CHE gene amplification more frequently occurs in such individuals because it provides the oocytes with advantageous properties to survival.

One wonders whether the pronounced expression of ChE genes in oocytes also implies that these genes are similarly expressed in developing sperm cells. It has long been known that cholinergic drugs affect sperm motility [292] and that ACh is present in mammalian spermatozoa [35, 36]. The correlation of ChEs with motility has also been proposed in relation with human respiratory epithelium ciliary cells [75]. More recently, AChE has been detected in sea urchin spermatozoa [59]. Further experiments will be required to examine whether the CHE and ACHE genes are expressed in developing human spermatozoa and to reveal their function in spermiogenesis.

6. Hereditary Defective CHE Gene Amplification: Putative Response to Organophosphorous Poisoning?

Three phenotypic variants for erythrocyte AChE were reported to reflect two codominant alleles at a single locus [72] which has never been genetically mapped. In avians as well, genetic information suggests that all

AChE forms in nerve and muscle are encoded by a single gene [287]. In contrast, serum BuChE displays multiple variants which have been genetically linked to two independent loci, CHE1 and CHE2. The E1 locus has genetically been assigned to a chromosome 3q region [14], in linkage with the transferrin (TF) gene, mapped at 3q26-q31, the transferrin receptor (TFRC) and the ceruloplasmin gene, with no further order of these four loci [reviewed in 174]. The CHE2 locus, directing the production of the common C5 variant of serum BuChE [305, 307, 347], has been a subject of contention. This argued locus has been shown to be genetically linked to the α-haptoglobin gene [211], however with a rather weak linkage. In situ hybridization experiments localized the haptoglobin-coding sequences to a region distal to the fragile site at 16q22. Our own gene-mapping attempts localized structural BuChE-coding sequences to two regions. These pointed at the presumptive CHE1 locus at 3q21-q26, proximal to the transferrin gene on the long arm of chromosome 3, and the CHE2 locus at a 16q11-q23 position, proximal to that of the haptoglobin gene on chromosome 16 [317, 359]. This analysis therefore revealed that sequences with the potential to encode BuChE, or with homology to such coding regions, do exist at least in some individuals at the long arm of chromosome 16.

Accumulated genetic evidence suggests that the CHE gene in the CHE1 locus is generally expressed, with nonfrequent CHE1 variants directing the production of defective ('silent' or 'atypical') enzyme [174, 305, 347]. In contrast, the CHE gene in the CHE2 locus leads to the expression of an active serum ChE in ca. 8% of the Western population only [211, 307]. Thus, the E2 gene might be a transcriptionally inactive pseudogene in the majority of individuals. Alternatively, frequent mutations in this gene could produce catalytically inactive forms of the serum BuChE protein. Further experiments will be needed to distinguish between these possibilities and determine whether any of these genes also codes, perhaps by alternate promoters or differential splicing, for other BuChEs, such as the membrane-associated BuChE in neuromuscular junctions [225] or soluble brain BuChE tetramers [357] or monomers [274]. In view of the single BuChE cDNA sequence that was so far found in all of our screening attempts, it seems highly likely that the CHE1 gene does code for all of the variable BuChE forms.

Apart from ACh hydrolysis, the roles of ChEs in general and of serum BuChE in particular are still unknown. Several clinical reports have linked cytogenetic anomalies in the 3q21-3q26 region in acute nonlymphocytic leukemia [32, 327] with thrombocythemia [58; reviewed in 320]. In view of the physiological effect of carbamylcholine in inducing megakaryocytopoiesis in culture [54, 258], as combined with the in situ hybridization localization of the CHE1 gene to this particular chromosomal region, it would be intriguing to propose that humoral ChE may take part in directing progenitor cells of the hemopoietic family to become committed promegakaryocytes. This suggestion is supported by the observation of CHE and ACHE gene amplifications in leukemias and platelet disorders [191].

Information accumulated by others, together with the findings presented in this monograph, point towards a pivotal direct or indirect role for members of the ChE gene family in cell division, growth and development in various biosystems, including bone marrow stem cells and germline cells. This could possibly indicate that in the case of the H family the amplification of the CHE gene might have given a growth advantage to the M.I. embryo, similar to amplified oncogenes in tumors. However, the nucleotide sequence of the BuChE cDNA does not resemble any of the known oncogenes or growth factors. The most likely explanation for the amplification event is, therefore, related with the ACh-hydrolyzing activity of BuChE and its putative implication in the biological control of cholinergic signalling.

This hypothesis is based on the assumptions that AChE and BuChE are interchangeable in many cell types playing the same catalytic function, and that BuChE in germline cells is essential for ACh hydrolysis, being the major ChE type in these cells. In an otherwise normal oocyte or sperm cell, overproduction of normal BuChE might interfere with cholinergic function and be lethal to the developing germline cell. In contrast, overexpressed defective BuChE would be less harmful due to its very low catalytic activity, while, perhaps, improving the resistance of a developing sperm cell, oocyte or embryo to OP poisoning. Multiple cases of chronic anti-ChE poisoning were reported in Israel [273], where the frequency of defective CHE phenotypes is particularly high [329]. We have found that both of M.I.'s parents were working in agriculture when M.I. was conceived, being exposed to high levels of parathion. When combined with the occurrence of a defective CHE gene, such exposure may have created conditions under which only the amplification and overproduction of such defective BuChE would permit survival. This event could be related to the extent of exposure, perhaps explaining why another sibling in the same family, also expressing the defective phenotype, does not carry the amplification. ChE genes are essential to the development of nematodes [78]. Houseflies resistant to OP insecticides [92] were found to produce different forms of insensitive AChE [93], perhaps suggesting that mutagenesis as well might provide OP resistance.

Once the H family was discovered, we were interested to find out whether the amplification of the human CHE gene is a unique phenomenon, particularly since OP poisons are continuously being exploited both as commonly used insecticides and as war agents [341]. The description of a Japanese family with heritable hypercholinesterasemia and isoenzymic alterations [149] suggested that other germline CHE gene amplications may yet be found. Moreover, in view of the overexpression of BuChE that we observed in brain tumors [274] and the altered properties of BuChE in the serum of carcinoma patients [358], it appeared possible that the CHE gene in somatic cells could also be subject to DNA amplification events. DNA transfection and transgenic mice experiments aimed to examine this point are the logical route for continuation of these experiments. In the future it would be important to determine whether repetitive exposure to OP poisons provides a selective pressure for gene amplification at the CHE locus and whether other commonly used chemical agents have similar effects on additional loci in man and other species.

7. Correlation of Cholinesterase Gene Amplification with Hemocytopoiesis in vivo

ChEs were proposed as markers for early cells of the megakaryocytic series already in 1973 [164]. To further search for CHE gene amplification, the postulated relationship between the family of ChEs and hematopoietic commitment and differentiation was investigated using cDNA probes. These probes detected the presence of multiple copies of the genes coding for AChE and BuChE in 25% of the leukemic DNA samples examined. Amplification of DNA sequences occurring at specific chromosomal breakpoints has been increasingly found in various malignancies [37]. In several cases, these changes were correlated with cellular growth and development effects [325]. One region that is conspicuously altered in leukemias appears on the long arm of chromosome No. 3 [32, 262, 327], where we recently mapped the CHE genes [317, 359]. Also, as mentioned above, ChE inhibitors and ACh analogs such as carbomylcholine induce abnormal proliferation of mouse megakaryocyte progenitor cells, both in vivo [54] and in vitro [55]. The augmented development of small AChE-positive cells in thrombocytopenic murine bone marrow cells has also been noted [356]. Taken together, this appeared to be sufficient to initiate a search for structural changes within the ACHE and CHE genes in leukemias. Our finding of 6 out of 20 amplification events among both genes, in cases of hematocytopoietic abnormalities, suggested that these apparently unrelated pieces of evidence might be connected.

The occurrence of these gene amplification events could reflect a specific origin of replication within the amplified CHE genes or in an adjacent, yet unknown oncogene [37, 325]. Yet another possibility is that of the insertion of a retroviral sequence, followed by the extension of its amplification into the chromosomal region of the ChE genes. The amplification of the CHE gene on chromosome No. 3 that we previously found in the H family, that was exposed to chronic doses of parathion [265], could be an example for the first option. It should be noted, however, that in that particular family the CHE gene was the only one to be amplified [Soreq et al., unpubl. observation]. Other examples are the changes in the Ig genes close to the c-myc oncogene in Burkitt's lymphoma [332], and the amplification of cellular DNA sequences at the boundaries of the insertion site of polyoma DNA [25].

The coamplification of the genes coding for both AChE and BuChE in our patients either indicates that these two genes were colocalized at the same chromosomal region prior to the amplification event and were amplified together, or reflects the occurrence of recombination events between the CHE and the ACHE genes during the amplification process. Alternatively, the ACHE and CHE genes might be independently subjected to the same selection pressure to be amplified. Amplification of various genes, including colocalized ones, has repeatedly been found in multidrug-resistant cell lines [31]. Chromosomal rearrangement has also been proposed to facilitate gene amplification in drug-resistant cells by juxtaposing homologous segments [124]. In the nematode *Caenorhabditis elegans*, multiple molecular forms of AChE [168] were associated with the appearance of specific AChE-deficient mutants [167, 170, 184, 185]. In *Drosophila*, a high frequency of novel recombinational events was noted for the *ace* locus, carrying the structural ACHE gene [242]. Precise mapping and phsyical linkage studies by pulse field gel electrophoresis, relating the yet unlocalized gene(s) coding for human AChE to those encoding BuChE, will be required to clarify this issue.

Appearance of novel restriction fragments in the pronounced cases of the ACHE and CHE gene amplifications could be due to overlapping, but nonequal regions of DNA, having been amplified in the various individuals, perhaps reflecting variable origins of replication resulting from retroviral transposition. Various insertion sites for amplifiable retroviral sequences have been observed in the human genome, including a chromosome 3q site for leukemia virus sequences [146] close to the location of the CHE genes [317]. Alternatively, the different patterns obtained in the various analyzed DNAs could reflect genetic alterations in the amplified genes, such as those observed for the amplified c-myc protooncogene in primary breast carcinomas [114] or those occurring in the dihydrofolate reductase (DHFR) gene in methotrexate-treated leukemic cells [293].

The possibility should be considered that the amplification of ChE-encoding genes was induced by continuous exposure to ChE inhibitors (i.e., agricultural OP insecticides, see Section III.6). The amplification of the ACHE and CHE genes in leukemias is not a random process, as it does not involve irrelevant sequences such as the ribosomal protein gene. ChE gene amplification could be advantageous to bone marrow stem cells to which ChE activities are essential by creating acquired resistance to ChE inhibitors, like the amplification and overexpression of multidrug resistance genes [296] and the amplification of genes induced by arsenic [200]. To further examine this possibility, the levels of expression of the amplified ACHE and CHE genes in hematocytopoietic disorders will have to be measured in individuals under chronic exposure to OP insecticides. Another approach to this issue would be to examine the direct effect of CHE genes inactivation on bone marrow cells differentiation.

The putative involvement of ChEs in the etiology of hematocytopoietic disorders is of particular importance, in view of the multiple indirect reports implicating these enzymes with growth and development [28, 121, 141, 188, 198, 199]. Assuming that ChEs are important for hematocytopoiesis, the amplification of ChE-encoding genes would be analogous to other amplifications in malignancies. Examples include that of the genes coding for the erb-B oncogene (mutated epidermal growth factor receptor) in malignant gliomas [203, 204, 351], the amplification of the *neu* oncogene in breast cancer, which is correlated with relapse and survival [289] and the amplification of *N-myc* in neuroblastoma, associated with the rate of progress of the disease [295]. Since AChE is also expressed in T cells [330] and in view of its altered regulation in erythrocyte defects such as paroxysmal nocturnal hemoglobinuria [69, 96], one may expect further findings of ChE gene amplification and/or mutagenesis in other blood cell disorders. Although ChEs are not homologous to oncogenes, there are indications for altered modes of their expression in ovarian carcinomas [100], malignant gliomas [274] and in the serum of patients with various carcinomas [358]. It would be interesting to reveal whether these reflect parallel amplification phenomena, giving multiple types of tumor cells growth advantages.

8. Modified Properties of Cholinesterases in the Serum of Carcinoma Patients Suggest that Antitumor Therapy Alters the Expression of Cholinesterase Gene(s)

Biochemical characterization of ChE properties in the serum of 77 patients suffering from primary carcinomas of different tissue origins, as compared with control serum of 21 healthy volunteers, revealed the appearance of a new type of ChE. The study was based on the analysis of a single serum sample from each patient, and in most cases serum was drawn after surgery from patients under surveillance and various treatment protocols according to the tissue origin of the tumor. The wide range of tumor types and wide ranges in the ages of these patients further complicate the interpretation of these findings. However, the reproducibility of inhibition patterns and gradient profiles, supported by the statistical analysis of a rather large group of samples, indicated that the modified properties of ChEs in the serum of these patients reflected a true in vivo phenomenon.

The cancer-associated soluble serum ChE was found to be susceptible to inhibition by both BW284C51 and iso-OMPA and exhibited a sedimentation coefficient of 6–7S in sucrose gradients. Thus its properties differed both from those of the well-characterized soluble serum BuChE, which is insensitive to BW284C51 and sediments as 12S tetramers [207], and from the properties of the dimeric erythrocyte AChE, which under normal conditions is not released to the serum in a soluble form and is not sensitive to iso-OMPA inhibition [147]. Partial proteolysis of various ChE forms does not alter their sensitivity to selective inhibitors [21, 252, 256, 263, 316, 317]. It is therefore unlikely that this tumor-characteristic type of serum ChE results from a disease-related release of AChE from the erythrocyte membranes or from breakdown of BuChE tetramers into dimers. This conclusion is further supported by the observation that BuChE dimers produced in microinjected oocytes are sensitive to iso-OMPA but not to BW284C51 [319]. Furthermore, the appearance of this new ChE type was not related to the average decrease in total serum ChE activity or to its general level of sensitivity to iso-OMPA and succinylcholine. This may imply that these two serum ChE activities originate from different pools of nascent polypeptides, perhaps by a yet unknown mode of alternative splicing. There is a single report in which the sensitivity of serum ChE to succinylcholine increased in a case of carcinoma [347]. However, in view of our present analysis, this seems to be an exception.

Could minor changes in the amino acid sequence of ChEs affect the sensitivity to selective inhibitors such as BW284C51? BW284C51-sensitive ChE has previously been detected by histochemical techniques in tissue sections derived from various types of primary carcinomas [100, 350]. In tissue homogenates from primary glioblastomas and meningiomas, we have found light forms of ChE which were sensitive to inhibition by both BW284C51 and iso-OMPA [274], however, it is not known yet whether these differences are the result of cell-type specific alternative splicing of mRNA transcripts like those of the calcitonin gene [282]. Microinjection of mRNA from gliobastomas, meningiomas and fetal brain into *Xenopus* oocytes induced the production of ChE activities which could be blocked by both inhibitors [312]. Altogether, this evidence supports the notion that the novel serum ChE fraction which was found in carcinoma patients was similar to the less characterized enzyme for which the term 'embryonic ChE' has been proposed [100]. It is possible that such enzyme is produced in embryonic tissues by a developmentally regulated mode of alternative splicing, parallel to similar mechanisms in other genes (i.e. CGRP) [9]. Such splicing may result in the formation of a distinct ChE type with yet undefined properties. Alternatively, the finding of biochemically altered ChE may reflect mutagenesis due to the antitumor therapy or within the tumor tissue which subscribes new and abnormal enzyme properties. A third possibility is that under chemotherapy, abnormal AChE-BuChE dimers are produced, similar to those in avian development [340]. However, it cannot be concluded at present whether this altered ChE is synthesized within and transported from the tumor tissue into the serum, or produced in another tissue and released as a response to the malignant state or to the treatment employed. To find out whether the production of embryonic ChE takes place within the tumor cells and is reduced after surgery, as is the case with many other tumor markers, longitudinal studies should be performed following the course of the disease in a carefully selected group of patients, all suffering from the same type of tumor, and treated and followed by the same protocol.

The primary structure of the soluble embryonic ChE and its relationship to other human ChEs, such as neuromuscular AChE or serum BuChE, are of particular interest. Understanding the regulation of this ChE may reveal the molecular control mechanisms leading to the tissue and cell type specificity of ChE polymorphism and shed light on the unknown physiological function of these carboxylesterases in proliferating and differentiating cells. The metabolism of AChE in cultured embryonic rat myotubes is affected by spontaneous electromechanical activity [48]. In pheochromocytoma cells [162], AChE biosynthesis is induced by nerve growth factor [46] and in glioblastomas enhanced levels of AChE [274] accompany the increase in epidermal growth factor receptor protein [203] and the amplification of the erb-B onocogene [204]. It would be interesting to examine whether the expression of ChE genes is coregulated with oncogenes and, if so, in what way. To approach this issue, ChE cDNA clones of primary tumor origin should be isolated. The nucleotide sequence of such tumor-originated ChE cDNA clones can be compared to the sequence of normal human BuChE cDNA and AChE cDNA from nonmalignant tissues [315, 318] to discover recurrent mutations. Should such mutations be found, they may serve as diagnostic tools to define the differentiation state of the examined tumors. This, in turn, could indicate the rate of progress and level of malignancy of such tumors. Oligodeoxynucleotide probes selective for such mutated sequences may subsequently be used as labelled probes to examine whether new types of ChE mRNA exist in carcinoma tissues.

9. Are Cholinesterases Involved with Regulation of Cellular Growth?

Several different modes of expression and gene amplifications were observed for ChEs in multiple types of developing and tumor human tissues. Taken together, these studies extend and support previous suggestions on the involvement of this family of proteins in cell division and/or growth mechanisms. In view of parallel studies on the amplification of genes producing target proteins to cytotoxic inhibitors, this raises the question whether OP poisons could be responsible for the selection pressure for CHE and ACHE gene amplifications.

Multiple behavioural effects of cholinesterase inhibitors have been known for a long time. These include general as well as highly specific effects, for example, alteration of receptive field properties in retina [15] or improvement of memory in the senile brain [29]. This implies a relationship with molecular processes affecting basic cellular mechanisms, such as cell division, DNA synthesis or mutagenesis. Once these are expected, it is not surprising to reveal the considerable body of information, accumulated over the years, which demonstrates cytotoxic and mitogenic effects for the commonly used OP insecticides. The threat of OP poisons to mammals in general, and more specifically to human kind, becomes readily apparent in view of the pesticide-induced DNA damage and perturbed repair processes that were observed by several research groups in cultured human cells [2, 4, 39]. These were directly related to chromosomal aberrations. Thus, sister chromatid exchange and cell cycle delay were demonstrated in cultured mammalian cells treated with many different OP insecticides [64]. These included the common agricultural insecticide malathion, which is very close in its structure to parathion [245]. Animal studies have, in parallel, shown similar effects in various species, including the mouse [89], guinea pig [94] and rat [224]. Damaging effects of OP poisons were also directly found in humans. First to be studied were high-risk groups including workers producing OP insecticides, who displayed transient chromosome aberrations [176]. Parallel chromosomal aberrations were also found in patients under acute organic phosphate insecticide intoxication [339], as well as in lymphocytes from agricultural workers during extensive occupational exposure to pesticides [355]. These observations did not attract much attention, perhaps because they were conceived as occupational risk of transient nature. However, it should be emphasized that the chromosomal damage induced by pesticides may not be limited to these high-risk groups of humans, since pesticide residues were found in basic food products [122]. This implies continuous subacute exposure to the entire population, with variable probabilities of clinical consequences.

Chromosome breakage occurring in germline cells, particularly sperm, is likely to induce hereditary changes in the affected genes. OP insecticides have been shown to inhibit testicular DNA synthesis [298] and induce sperm abnormalities in mice [353]. The implications of these findings may explain the hereditary CHE gene amplification in the parathion-exposed H family [265], in view of their defective CHE phenotype. Animal studies have also demonstrated that continuous administration of methyl parathion suppressed growth and induced ossification in both mice and rats. Further embryonic effects included high mortality and cleft palate in the mouse [reviewed in 320]. In humans, we found one paper reporting malformations of the extremities and fetal death correlated with exposure to methyl parathion in 18 cases [248]. In addition, a neonatal lethal syndrome of multiple malformations was more recently reported in women exposed to unspecified insecticides during early pregnancy [151]. All of these findings are most probably related with the developmental function(s) of ChEs being impaired by these poisons, and may possibly reflect CHE and ACHE gene mutagenesis and/or amplifications as well as chromosome breakage in the affected individuals.

Assuming that a causal relationship may exist between the exposure to OP insecticides and CHE/ACHE gene amplifications, a new fundamental importance should be postulated for cholinergic responses in cell division and growth, reproduction and embryogenesis. This assumption is supported by findings demonstrating the stimulation of DNA synthesis in brain-derived cells exposed to ACh analogues [16]. These experiments were performed using glial cell cultures transfected with the muscarinic ACh receptor genes. In these cells, cholinergic response depended on the expression of specific muscarinic receptor subtypes and operated via the hydrolysis of phosphatidylinositol. One could postulate that similar induction of DNA synthesis may occur in other cell types under exposure to agricultural OP insecticides. These block ACh hydrolysis by covalently interacting with ChEs and thus induce increased concentrations of the G protein-linked neurotransmitter ACh. Treatment of OP poisoning patients with purified serum BuChE was found to be effective [177]. One natural way to block OP effects on nuclear functioning would be the amplification of ChE genes. Figure 18 presents a scheme of a putative feedback mechanism which links CHE gene amplification to OP intoxication.

It has previously been shown that oocytes from various species are sensitive to ACh, and that ACh analogues induce phosphatidylinositol hydrolysis in *Xenopus* oocytes [reviewed in 297]. As mentioned above, ACh analogues and ChE inhibitors also induce promegakaryocytopoiesis in mouse bone marrow cultures [reviewed in 320]. Cholinergic intercellular communication may hence play pivotal role(s) in developing cells in

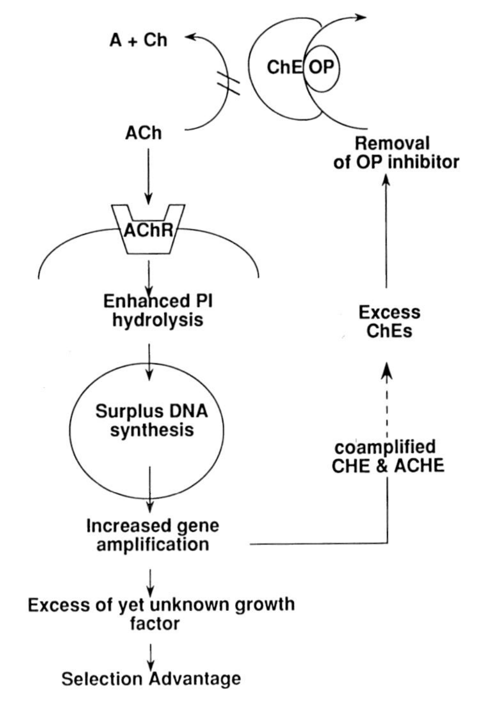

Fig. 18. A suggested feedback mechanism between OP intoxication and ChE gene amplification.

general, in addition to its important function in communicating between cholinergic neurons and in neuromuscular junctions.

OP poisons block the catalytic activities of ChEs [181] interfering with ACh hydrolysis and disturbing transport of ions through ACh receptor channels. Such disturbance has been shown to induce the enhanced hydrolysis of phosphatidylinositols in cultured cells and result in surplus DNA synthesis [16]. Surplus DNA synthesis increases the statistical chances for the amplification of genes which include origins of replication [325]. In cases where amplified CHE/ACHE genes produce excess of their protein products, OP inhibition may be overcome. This would provide cells in which such amplification occurred with selection advantage over other cells dependent on cholinergic signalling under OP exposure. While this complete circle may explain the amplification of ChE genes in germline cells, it does not fully relate to their amplification in tumors. Hence, one should expect that in tumor cells the same process should also result in an increase in a yet unknown growth factor or cell division-related gene, the enhanced expression of which is directly related with the transcriptional activity of the CHE locus.

10. Prospects for Future Research

Much of the progress in recent ChE molecular biology has been achieved through the isolation and characterization of AChE and BuChE cDNA and genomic clones from various species. The cloning of ChE genes from *Torpedo* electric organ, *Drosophila*, and man has facilitated sequence homology studies, chromosome mapping analyses, and the characterization of gene expression and mRNA processing events in various tissue types. This progress includes studies of the human ACHE gene and its protein product, AChE. Further progress must be complemented by efforts to isolate the human genomic CHE and ACHE promoters to facilitate analyses of the complex patterns of expression and regulation of these enzymes in man. Furthermore, completion of this work will open the door for large-scale production of recombinant AChE. Should this goal be achieved, the possibility of employing administered doses of purified AChE as a prophylactic treatment of OP poisoning will become a testable reality.

The finding of tumor-characteristic molecular forms of ChE in primary brain tumors [274] and carcinomas [358] raises the possibility that ChE-encoding genes are frequently subject to mutations or rearrangements, similarly to the gene coding for EGF receptor in such tumors [204]. Another question of particular interest in the case of the human ChE genes is in what way the tissue-specific expression of these genes derives from differential splicing, alternate transcription, the expression of different but related genes and/or other events, such as posttranslational processing and specific protein transport and degradation mechanisms. In the future, one should hope that research efforts will be focused on these issues.

The role of ChEs in nonnervous tissues poses an intriguing puzzle. Evidence is accumulating to support the notion that ChEs fulfill a function in developing cell types. Thus, continued exploration of the occurrence and role of normal and abnormal ChE expression in nondifferentiated cell types, and in neoplastic tissues of various types and stages of development, is essential. The correlation of ACHE and CHE gene amplifications with disease states, and its possible involvement in tumor etiology, requires further study. The possible correlations between cancer, CHE/ACHE gene amplifications, and sublethal OP exposure has far-reaching implications for human health which must be clarified.

Is the amplification of ChE genes limited to cells or organisms exposed to ChE inhibitors, or does it represent a normal process? In addition to the amplification of genes encoding target proteins for cytotoxic inhibitors and to the amplification of oncogenes enhancing growth and cell division processes, particular types of gene amplification also appear to represent normal processes in the development of various organisms. A well-known example is the transient amplification of chorion genes in insect larvae [323]. To the best of our knowledge, there are as yet no parallel examples of transiently amplified genes during mammalian development. However, the amplification and overexpression of the N-myc oncogene in primary neuroblastoma tumors has been proposed to represent a normal embryonic process, reflecting the undifferentiated nature of these tumor cells [179]. It would be interesting to examine whether the ChE genes are normally subjected to embryonic amplification during germline cells development and fetal development.

How amplified and overexpressed CHE and ACHE genes might enhance the rate of tumor cell growth and division is a yet unexplained mechanism. First, it should cause a considerable decrease in the concentration of available ACh. This, in turn, could induce changes in DNA synthesis, as implied from the work of Ashkenazi et al. [16]. It is possible that severe changes in nuclear functioning would then trigger the amplification of other genes including origins of replication, for example, various oncogenes. In this case, the phenomenon of ChE gene amplification should be related with relapse and/or progress of disease and could be considered as a prognostic/diagnostic measure or as a follow-up marker for particular tumors. Such measures would become particularly valuable if, as is implied from our work, the amplification of the CHE and the ACHE genes precedes that of oncogenes in various tumors. Further future directions for the investigation of ChE gene amplification should hence be focused on efforts to block the expression of BuChE mRNA and AChE mRNA transcripts. This could be done, perhaps, by 'antisense' oligodeoxynucleotides [360]. This novel approach may lead to the development of an efficient therapeutic treatment that would block the overproduction of ChEs and thus prevent their growth-related effects. Several other directions exist for studying the amplification of human ChE genes, an intriguing phenomenon with multiple basis and applied implications, and will largely depend on future observations.

References

1 Adams RD, Victor M: Principles of clinical myology: diagnosis and classification of muscle diseases; in Adams RD, Victor M (eds): Principles of Neurology. New York, McGraw-Hill, 1981, pp 940–949.

2 Ahmed F, Hart R, Lewis N: Pesticide induced DNA damage and its repair in cultured human cells. Mutat Res 1977;42:161–174.

3 Ajmar F, Garre C, Sessarego M, Ravazzolo R, Barresi R, Bianchi-Scarra G, Lituania M: Expression of erythroid acetylcholinesterase in the K-562 leukemia cell line. Cancer Res 1983;43:5560–5563.

4 Alam M, Kasatiya S: Cytological effects of an organic phosphate pesticide on human cells in vitro. Can J Cytol 1976;18:665–672.

5 Aldridge WN, Reiner E: Enzyme inhibitors as substrates. Amsterdam, North-Holland, 1972.

6 Alexandrova M, Holzbauer M, Racke K, Sharman DF: Acetylcholinesterase in the rat neurohypophysis is decreased after dehydration and released by stimulation of the pituitary stalk. Neuroscience 1987;21:421–427.

7 Allderdice PW, Browne N, Murphy DP: Chromosome 3 duplication q21 → qter deletion p25 → pter syndrome in children of carriers of a pericentric inversion inv (3) (p25q21). Am J Hum Genet 1975;27:699–718.

8 Alles GA, Hawes RC: Cholinesterases in the blood of man. J Biol Chem 1940;133:375–390.

9 Amara SG, Arriza JL, Leff SE, Swanson LW, Evans RM, Rosenfeld MG: Neuropeptide homologous to calcitonin-gene related peptide. Science 1985;229:1094–1097.

10 Anderson MJ, Fambrough DM: Aggregates of acetylcholine receptors are associated with plaques of a basal lamina heparan sulfate proteoglycan on the surface of skeletal muscle fibers. J Cell Biol 1983;97:1396–1411.

11 Anglister L, Silman I: Molecular structure of elongated forms of electric eel acetylcholinesterase. J Mol Biol 1978;125:293–311.

12 Anglister L, McMahan UJ: Basal lamina directs AChE accumulation at synaptic sites in regenerating muscle. J Cell Biol 1985;101:735–743.

13 Appleyard ME, Smith AD, Wilcock GK, Esiri MM: Decreased CSF acetylcholinesterase activity in Alzheimer's disease. Lancet 1983;ii:452.

14 Arias S, Rolo M, Gonzalez N: Gene dosage effect present in trisomy 3q25.2-qter for serum cholinesterase (CHE1) and absent for transferrin (TF) and ceruloplasmin (CP). Cytogenet Cell Genet 1985;40:571.

15 Ariel M, Daw NW: Effects of cholinergic drugs on receptive field properties of rabbit retinal ganglion cells. J Physiol 1982;344:135–160.

16 Ashkenazi A, Ranmachandran J, Capon DJ: Acetylcholine analogue stimulates DNA synthesis in brain-derived cells via specific muscarinic receptor subtypes. Nature 1989;340:146–150.

17 Astrin KH, Desnick RJ, Smith M: Provisional assignment of esterase B3 (ESB3) to human chromosome 16. Cytogenet Cell Genet 1982;32:250.
18 Atack JR, Perry EK, Bonham JR, Perry RH, Tomlinson BE, Blessed G, Fairbairn A: Molecular forms of acetylcholinesterase in senile dementia of Alzheimer type: Selective loss of the intermediate (10S) form. Neurosci Lett 1983;40:199–204.
19 Atack JR, Perry EK, Bonham JR, Candy JM, Perry RH: Molecular forms of acetylcholinesterase and butyrylcholinesterase in the aged human central nervous system. J Neurochem 1986;47:263–277.
20 Augustinsson KB: The evolution of esterases in vertebrates; in Their NV, Roche J (eds): Homologous Enzymes and Biochemical Evolution. New York, Gordon & Breach, 1968, pp 299–311.
21 Austin L, Berry WK: Two selective inhibitors of cholinesterase. Biochem J 1953;54:695–700.
22 Ayalon A, Zakut H, Prody CA, Soreq H: Preferential transcription of acetylcholinesterase over butyrylcholinesterase mRNAs in fetal human cholinergic neurons; in Giuffrida-Stella AM (eds): Gene Expression in the Nervous System. New York, Liss, in press.
23 Balasubramanian AS: Have cholinesterases more than one function? Trends Biochem Sci 1984;9:467–468.
24 Barakat I, Massarelli R, Courageot J, Devilliers G, Sensenbrenner M: Development of cholinergic properties in nerve cell cultures in the presence of brain extract. Brain Res 1983;279:207–216.
25 Baran N, Lapidot A, Manor H: Amplification of cellular DNA sequences at the boundaries of the insertion site of polyoma DNA. Mol Cell Biol 1987;7:2636–2640.
26 Barnard EA, Miledi R, Sumikawa K: Translation of exogenous messenger RNA coding for nicotinic acetylcholine receptors produces functional receptors in *Xenopus* oocytes. Proc R Soc Lond B 1982;215:241–248.
27 Barnard EA: Multiple molecular forms of acetylcholinesterase and their relationship to muscle function; in Brzin M, Barnard EA, Sket D (eds): Cholinesterases: Fundamental and Applied Aspects. Proc 2nd Int Meet on Cholinesterases, Bled, Yugoslavia. Berlin, de Gruyter, 1984, pp 49–71.
28 Bartos EM, Glinos AD: Properties of growth-related acetylcholinesterase in a cell line of fibroblastic origin. J Cell Biol 1976;69:638–646.
29 Bartus RT, Dean RL, Beer B, Lippa AS: The cholinergic hypothesis of geriatric memory dysfunction. Science 1982;217:408–417.
30 Bause E: Structural requirements of N-glycosylation of proteins. Biochem J 1983;209:331–336.
31 Beidler JL, Chang TD, Scotto KW, Melera PW, Spengler BA: Chromosomal organization of amplified genes in multidrug-resistant Chinese hamster cells. Cancer Res 1988;48:3179–3187.
32 Bernstein R, Pinto MR, Behr A, Mendelow B: Chromosome abnormalities in acute nonlymphocytic leukemia (ANLL) with abnormal thrombopoiesis: Report of three patients with a 'new' inversion anomaly and a further case of homologous translocation. Blood 1982;60:613–617.
33 Bianchi Scarra GL, Garre C, Ravazzolo R, Coviello D, Origoni P: Coordinated expression of acetylcholinesterase and hemoglobin in K562 cells induced to terminal differentiation by cytosine arabinoside (ara-c). Cell Biol Int Rep 1986;10:167.
34 Bick DP, Balkite EA, Baumgarten A, Hobbins JC, Mahoney MJ: The association of congenital skin disorders with acetylcholinesterase in amniotic fluid. Prenat Diagn 1987;7:543–549.

35 Bishop MR, Sastry BVR, Schmidt DE, Harbison RD: Spermic cholinergic system and occurrence of acetylcholine and other quarternary ammonium compounds in mammalian spermatozoa. Toxicol Appl Pharmacol 1975;33:733–734.

36 Bishop MR, Sastry BVR, Schmidt DE, Harbison RD: Occurrence of choline acetyltransferase and acetylcholine and other quaternary ammonium compounds in mammalian spermatozoa. Biochem Pharmacol 1976;25:1617–1622.

37 Bishop JM: The molecular genetics of cancer. Science 1987;235:305–311.

38 Bisso GM, Meneguz A, Michalek H: Developmental factors affecting brain acetylcholinesterase inhibition and recovery in DFP-treated rats. Dev Neurosci 1982;5:508–519.

39 Blevins R, Lijinsky W, Regan JM: Nitrosated methylcarbamate insecticides: effect on the DNA of human cells. Mutat Res 1977;44:1–7.

40 Bonham JR, Atack JR: A neural tube defect specific form of acetylcholinesterase in amniotic fluid. Clin Chem Acta 1983;135:233–237.

41 Bradford MM: A rapid and sensitive method for the quantitation of microgram quantities of protein utilizing the principle of protein-dye binding. Anal Biochem 1978;72:248–253.

42 Brandan E, Inestrosa NC: The synaptic forms of acetylcholinesterase bind to cell-surface heparan sulfate proteoglycans. J Neurosci Res 1986;15:185–196.

43 Branks PL, Wilson MC: Patterns of gene expression in the murine brain revealed by in situ hybridization of brain-specific mRNAs. Mol Brain Res 1986;1:1–16.

44 Brezenoff HE, McGee J, Hymowitz N: Inhibition of acetylcholinesterase in the gut inhibits schedule-controlled behavior in the rat. Life Sci 1985;37:49–54.

45 Brimijoin S, Mintz KP, Alley MC: Production and characterization of separate monoclonal antibodies to human acetylcholinesterase and butyrylcholinesterase. Mol Pharmacol 1983;24:513–520.

46 Brimijoin S, Rakonczay Z: Immunology and molecular biology of the cholinesterases: Current results and prospects. Int Rev Neurobiol 1986;28:363–410.

47 Brock DJH, Bader P: The use of commercial antisera in resolving the cholinesterase bands of human amniotic fluids. Clin Chem Acta 1983;127:419–422.

48 Brockman SK, Younkin LH, Younkin SG: The effect of spontaneous electromechanical activity on the metabolism of acetylcholinesterase in cultured embryonic rat myotubes. J Neurosci 1984;4:131–140.

49 Brockman SK, Usiak MF, Younkin SG: Assembly of monomeric acetylcholinesterase into tetrameric and asymmetric forms. J Biol Chem 1986;261:1201–1207.

50 Brockman SK, Younkin SG: Effect of fibrillation on the secretion of acetylcholinesterase from cultured embryonic rat myotubes. Brain Res 1986;376:409–411.

51 Brown T, Kennard O, Kneale G, Rabinovitch D: High-resolution structure of a DNA helix containing mismatched base pairs. Nature 1985;315:604–606.

52 Bull D: The Growing Problem: Pesticides and the Third World, Oxford, Oxfam, 1982.

53 Burmeister N, Avivi A, Schlessinger J, Soreq H: Production of EGF-containing polypeptides in *Xenopus* oocytes microinjected with submaxillary gland mRNA. EMBO J 1984;3:1499–1505.

54 Burstein SA, Adamson JW, Harker LA: Megakaryocytopoiesis in culture: Modulation by cholinergic mechanisms. J Cell Physiol 1980;103:201–208.

55 Burstein SA, Harker LA: Control of platelet production. Clin Haematol 1983;12:3–27.

56 Burstein SA, Boyd CN, Dale GL: Quantitation of megakaryocytopoiesis in liquid culture by enzymatic determination of acetylcholinesterase. J Cell Physiol 1985;122:159–165.

57 Caras IW, Weddell GN: Signal peptide for protein secretion directing glycophospholipid membrane anchor attachment. Science 1989;243:1196–1198.

58 Carbonell F, Hoelzer D, Thiel E, Bartl R: Ph-1 positive CML associated with megakaryocytic hyperplasia and thrombocytemia and an abnormality of chromosome 3. Cancer Genet Cytogenet 1982;6:153–161.
59 Cariello L, Romano G, Nelson L: Acetylcholinesterase in sea urchin spermatozoa. Gamete Res 1986;14:323–332.
60 Carter P, Wells JA: Engineering enzyme specificity by 'substrate-assisted catalysis'. Science 1987;237:394–399.
61 Catalan RE, Aragones MD, Godoy JE, Martinez AM: Ecdysterone induces acetylcholinesterase in mammalian brain. Comp Biochem Physiol 1984;78C:193–195.
62 Chao LP, Kan KKS, Angelini C, Keesey J: Autoimmune neuromuscular disease induced by a preparation of choline acetyltransferase. Exp Neurol 1982;75:23.
63 Chatonnet A, Lockridge O: Comparison of butyrylcholinesterase and acetylcholinesterase. Biochem J 1989;260:625–634.
64 Chen HR, Hsueh J, Sirlanni S, Huang C: Induction of sister chromatid exchanges and cell cycle delay in cultured mammalian cells treated with eight organophosphorous insecticides. Mutat Res 1981;88:307–316.
65 Chen HR, Dayhoff MO, Barker WC, Hunt LT, Yen LS, George DG, Orcutt BC: Nucleic acid sequence data-base. IV. DNA 1982;1:365–374.
66 Cherbas P, Cherbas L, Williams CM: Induction of acetylcholinesterase activity by beta-ecdyson in a *Drosophila* cell line. Science 1977;197:275–277.
67 Chou PY, Fasman GD: A prediction of secondary structure of proteins from their amino acid sequence. Adv Enzymol 1978;47:45–147.
68 Chou PY, Fasman GD: Empirical predictions of protein conformations. Annu Rev Biochem 1978;47:251–276.
69 Chow FL, Telen MJ, Rosse WF: The acetylcholinesterase defect in paroxysmal nocturnal hemoglobinuria: Evidence that the enzyme is absent from the cell membrane. Blood 1985;66:940–945.
70 Chubb W, Goodman S, Smith A: Is acetylcholinesterase secreted from central neurons into the cerebro-spinal fluid? Neuroscience 1976;1:57–62.
71 Chubb IW, Bornstein JC: Dopamine and acetylcholinesterase release in the substantia nigra: Cooperative or coincidental? Neurochem Int 1985;7:905–912.
72 Coates PM, Simpson NE: Genetic variation in human erythrocyte acetylcholinesterase. Science 1972;175:1466–1477.
73 Coggin JH: The implications of embryonic gene expression in neoplasia. CRC Crit Rev Oncol Hematol 1986;5:37–55.
74 Cooper GW, Cooper B: Relationships between blood platelet and erythrocyte formation. Life Sci 1977;20:1571–1580.
75 Cossen G, Allen CR: Acetylcholine; its significance in controlling ciliary activity of human respiratory epithelium in vitro. J Appl Physiol 1959;14:901–904.
76 Cox KH, DeLeon DV, Angerer LM, Angerer RC: Detection of mRNAs in sea urchin embryos by in situ hybridization using asymmetric RNA probes. Dev Biol 1984;101:485–502.
77 Coyle JT, Price DL, DeLong MR: Alzheimer's disease: a disorder of cortical cholinergic innervation. Science 1983;219:1184–1190.
78 Culotti JC, Von Ehrenstein G, Culotti MR, Russell RL: A second class of acetylcholinesterase-deficient mutants of the nematode *Caenorhabditis elegans*. Genetics 1981;97:281–305.
79 Couteaux R, Taxi J: Recherches histochimiques sur la distribution des activités cholinestératiques an niveau de la synapse myoneurale. Arch Anat Microscop Morphol Exp 1952;41:352–392.

80 Couteaux R: Localization of cholinesterases at neuromuscular junctions. Int Rev Cytol 1955;4:335–375.

81 Craik CS, Roczniak S. Largman C, Rutter WJ: The catalytic role of active site aspartic acid in serine protease. Science 1987;237:909–913.

82 Danilchik MV, Gerhardt JC: Differentiation of the animal-vegetal axis in *Xenopus laevis* oocytes. Dev Biol 1987;122:101–112.

83 Dascal N, Landau EM: Types of muscarinic response in *Xenopus* oocytes. Life Sci 1980;27:142–146.

84 Dascal N, Landau EM, Lass Y: Acetylcholine promotes progesterone induced maturation of *Xenopus* oocytes. J Physiol (Lond) 1984;352:551–574.

85 Dascal N, Snutch TP, Lubbert H, Davidson N, Lester HA: Expression and modulation of voltage gated calcium channels after RNA injection in *Xenopus* oocytes. Science 1986;231:1174–1150.

86 Davies RO, Marton AV, Kalow W: The action of normal and atypical cholinesterase of human serum upon a series of esters of choline. Can J Biochem Physiol 1960;38:545–551.

87 Dayhoff MO, Barker WC, Hunt LT: Establishing homologies in protein sequences. Methods Enzymol 1983;91:524–545.

88 Dayhoff MO: Atlas of Protein Sequence and Structure. Washington, National Biomedical Research Foundation, 1985, suppl 3, p 5.

89 Degraeve N, Moutschen J: Genotoxicity of an organophosphorous insecticide, dimethoate, in the mouse. Mutat Res 1983;119:331–337.

90 Delfs JR, Zhu CH, Dichter MA: Coexistence of acetylcholinesterase and somatostatin-immunoreactivity in neurons cultured from rat cerebrum. Science 1984;223:61–63.

91 Devereux J, Haeberli P, Smithies O: A comprehensive set of sequence analysis programs for the VAX. Nucleic Acids Res 1984;12:387–395.

92 Devonshire AL: Studies of the acetylcholinesterase from houseflies (*Musca domestica* L.) resistant and susceptible to organophosphorous insecticides. Biochem J 1975;149:463–468.

93 Devonshire AL, Moores GD: Different forms of insensitive acetylcholinesterase in insecticide-resistant houseflies (*Musca domestica*). Pest Biochem Physiol 1984;21:336–340.

94 Dikshith T: In vivo effects of parathion on guinea pig chromosomes. Environ Physiol Biochem 1973;3:161–168.

95 Dobkin BH, Verity MA: Familial neuromuscular disease with type 1 fiber hypoplasia, tubular aggregates, cardiomyopathy and myasthenic features. Neurology 1978;28:1135.

96 Dockter ME, Morrison M: Paroxysmal nocturnal hemoglobinuria erythrocytes are of two distinct types: Positive or negative for acetylcholinesterase. Blood 1986;67:540–543.

97 Doctor BP, Camp S, Gentry MK, Taylor SS, Taylor P: Antigenic and structural differences in the catalytic subunits of the molecular forms of acetylcholinesterase. Proc Natl Acad Sci USA 1983;80:5767–5771.

98 Doctor BP, Toker L, Roth E, Silman I: Microtiter assay for acetylcholinesterase. Anal Biochem 1987;166:399–403.

99 Doolitle RF: The proteins. Sci Am 1985;253:74–83.

100 Drews E: Cholinesterase in embryonic development. Prog Histochem Cytochem 1975;7:1–52.

101 Dreyfus PA, Rieger F, Pincon-Raymond M: Acetylcholinesterase of mammalian neuromuscular junctions: presence of tailed asymmetric acetylcholinesterase in synaptic basal lamina and sarcolemma. Proc Natl Acad Sci USA 1983;80:6698–6702.

102 Dreyfus PA, Friboulet A, Tran LH, Rieger F: Polymorphism of acetylcholinesterase and

identification of new molecular forms after sedimentation analysis. Biol Cell 1984;51:35–42.

103 Dreyfus PA, Verdiere M, Goudou D, Garcia L, Rieger F: Acetylcholinesterase in mammalian skeletal muscle and sympathetic ganglion cells. Extra- and intracellular hydrophilic and hydrophobic asymmetric forms; in Changeux, Hucho, Maelicke, Neuman (eds): Molecular basis of nerve activity. Berlin, de Gruyter, 1985, pp 729–739.

104 Dreyfus PA, Zevin-Sonkin D, Seidman S, Prody C, Zisling R, Zakut H, Soreq H: Cross-homologies and structural differences between human cholinesterases revealed by antibodies against cDNA-produced butyrylcholinesterase peptides. J Neurochem 1988;51:1858–1867.

105 Dreyfus PA, Seidman S, Pincon-Raymond M, Murawsky M, Rieger F, Schejter E, Zakut H, Soreq H: Tissue-specific processing and polarized compartmentalization of clone-produced cholinesterase in microinjected *Xenopus* oocytes. Mol Cell Neurobiol 1989;9:323–341.

106 Dreyfus PA: Multileveled regulation of the human cholinesterase genes and their protein products; PhD thesis, The Hebrew University of Jerusalem, 1989.

107 Dumont JN: Oogenesis in *Xenopus laevis* (Daudin). 1. Stages of oocyte development in laboratory-maintained animals. J Morphol 1972;136:153–180.

108 Dutta-Choudhury TA, Rosenberry T: Human erythrocyte acetylcholinesterase is an amphipathic protein whose short membrane-binding domain is removed by papain digestion. J Biol Chem 1984;259:5653–5660.

109 Dziegielewska KM, Saunders NR, Soreq H: Messenger ribonucleic acid (mRNA) from developing rat cerebellum directs in vitro biosynthesis of plasma proteins. Dev/Brain Res 1985;23:259–267.

110 Dziegielewska KM, Saunders NR, Schejter EJ, Zakut H, Zevin-Sonkin D, Zisling R, Soreq H: Synthesis of plasma proteins in fetal, adult and neoplastic human brain tissue. Dev Biol 1986;115:93–104.

111 Edwards RG: Chromosomal abnormalities in human embryos. Nature 1983;303:283–286.

112 Ellman GL, Courtney DK, Anders V, Featherstone RM: A new and rapid colorimetric determination of acetylcholinesterase activity. Biochem Pharmacol 1961;7:88.

113 Elmquist D, Lambert E: Detailed analysis of neuromuscular transmission in patients with the myasthenic syndrome sometimes associated with bronchogenic carcinoma. Mayo Clin Proc 1986;43:681.

114 Escot C, Theillet C, Lidereau R, Spyratos F, Champeme MH, Crest J, Callahan R: Genetic alteration of the c-myc protooncogene (MYC) in human primary breast carcinoma. Proc Natl Acad Sci USA 1986;83:4834–4838.

115 Eusebi F, Mangia F, Alfei L: Acetylcholine-elicited responses in primary and secondary mammalian oocytes disappear after fertilization. Nature 1979;277:651–653.

116 Eusebi F, Pasetto N, Siracusa G: Acetylcholine receptors in human oocytes. J Physiol (Lond) 1984;346:321–330.

117 Evans DH, Morgan AR: Characterization of imperfect DNA duplexes containing unpaired bases and non-Watson-Crick base pairs. Nucleic Acids Res 1986;14:4267–4280.

118 Faff J, Bak W: Adaptation of the neuromuscular junction to chronic acetylcholinesterase inhibition due to phospholine treatment. Arch Int Pharmacodyn 1981;250:293.

119 Fambrough DM, Engel AG, Rosenberry TL: Acetylcholinesterase of human erythrocytes and neuromuscular junctions: Homologies revealed by monoclonal antibodies. Proc Natl Acad Sci USA 1982;79:1078–1082.

120 Festoff BW, Israel RS, Engel WK, Rosenbaum RB: Neuromuscular blockade with anti-exoplasmic antibodies. Neurology 1977;27:963.

121 Fitzpatrick-McElligot S, Stent GS: Appearance and localization of acetylcholinesterase in embryo of the leech *Helobdella triserialis*. J Neurosci 1981;1:901–907.

122 Food and Agriculture Organization of the United States (FAO): Pesticide residues in food. FAO Plant Production and Protection Paper IS Rev. 1979;27:379–392.

123 Feinberg AP, Vogelstein B: A technique for radiolabeling DNA restriction endonuclease fragments to high specific activity. Anal Biochem 1983;132:6–13.

124 Flintoff WF, Livingston E, Duff C, Warton RG: Moderate-level gene amplification in methotrexate-resistant Chinese hamster ovary cells is accompanied by chromosomal translocations at or near the site of the amplified DHFR gene. Mol Cell Biol 1984;4:69–76.

125 Flormann HM, Storey BT: Inhibition of in vitro fertilization of mouse eggs: 3-quincyclidinylbenzylate specifically blocks penetration of zonae pellucidae by mouse spermatozoa. J Exp Zool 1981;216:159–167.

126 Fossier P, Baux G, Tauc L: Possible role of acetylcholinestesterase in regulation of postsynaptic receptor efficiency at a central inhibitory synapse of aplysia. Nature 1983;301:710–712.

127 Fritze J, Beckmann H: Erythrocyte acetylocholinesterase in psychiatric disorders and controls. Biol Psychiat 1987;22:1097–1106.

128 Fuller SD, Bravo R, Simons K: An enzymatic assay reveals that proteins destined for the apical or basolateral domains of an epithelial cell line share the same late Golgi apparatus. EMBO J 1985;4:297–307.

129 Futerman AH, Low MG, Silman I: A hydrophobic dimer of acetylcholinesterase from *Torpedo californica* electric organ is solubilized by phosphatidylinositol-specific phospholipase C. Neurosci Lett 1983;40:85–89.

130 Futerman AH, Low MG, Michaelson DM, Silman I: Solubilization of membrane-bound acetylcholinesterase by a phosphatidylinositol-specific phospholipase C. J Neurochem 1985;45:1487–1494.

131 Garcia L, Verdiere-Sahuque M, Dreyfus PA, Nicolet M, Rieger F: Association of tailed acetylcholinesterase to lipidic structures in mammalian skeletal muscle. Neurochem Int 1988;13:231–236.

132 Garnier J, Osgathorpe DJ, Rovson B: Analysis of the accuracy and implications of simple methods for predicting the secondary structure of globular proteins. J Mol Biol 1978;120:97–120.

133 Gennari K, Brodbeck U: Molecular forms of acetylcholinesterase from human caudate nucleus: Comparison of salt-soluble and tetrameric enzyme species. J Neurochem 1985;44:697–704.

134 Gennari K, Brunner J, Brodbeck U: Tetrameric detergent-soluble acetylcholinesterase from human caudate nucleus: Subunit composition and number of active sites. J Neurochem 1987;49:12–18.

135 Gibney G, MacPhee-Quigley K, Thompson B, Vedvick T, Low MG, Taylor SS, Taylor P: Divergence in primary structure between the molecular forms of acetylcholinesterase. J Biol Chem 1988;263:1140–1145.

136 Gibney B, MacPhee-Quigley K, Maulet Y, Schumacher M, Camp S, Taylor P: Alternative mRNA splicing gives rise to asymmetric and glycophospholipid linked forms of acetylcholinesterase (abstract). FASEB, 1988.

137 Gibson GE, Peterson C, Jenden DJ: Brain acetylcholine synthesis declines with senescence. Science 1981;213:674–676.

138 Grassi J, Vigny M, Massoulie J: Molecular forms of acetylcholinesterase in bovine

caudate nucleus and superior cervical ganglion: Solubility properties and hydrophobic character. J Neurochem 1982;38:457–469.
139 Graybiel AM, Picke VM, Joh TH, Reis DJ, Ragsdale CW: Direct demonstration of a correspondence between the dopamine islands and acetylcholinesterase patches in the developing striatum. Proc Natl Acad Sci USA 1981;78:5871–5875.
140 Graybiel AM, Ragsdale CW, Yoneoka ES, Elde RP: An immunohistochemical study of enkephalins and other neuropeptides in the striatum of cat with evidence that the opiate peptides are arranged to form mosaic patterns in register with the striosomal compartments visible by acetylcholinesterase staining. Neurosci 1981;6:377–397.
141 Greene LA, Rukenstein A: Regulation of acetylcholinesterase activity by nerve growth factor. Role of transcription and dissociation from effects on proliferation and neurite outgrowth. J Biol Chem 1981;256:6363–6367.
142 Greenfield SA, Smith AD: The influence of electrical stimulation of certain brain regions on the concentration of acetylcholinesterase in rabbit cerebrospinal fluid. Brain Res 1979;177:445–459.
143 Greenfield S, Cheramy A, Leviel V, Glowinski J: In vivo release of acetylcholinesterase in cat substantia nigra and caudate nucleus. Nature 1980;284:355–357.
144 Greenfield S: Acetylcholinesterase may have novel functions in the brain. Trends Neurosci 1984;7:364–368.
145 Gunderson CB, Miledi R: Acetylcholinesterase activity of *Xenopus laevis* oocytes. Neuroscience 1985;10:1487–1495.
146 Gusella JF, Wasmuth JJ: Human Gene Mapping 9: Report of the Committee on Chromosomes 3 and 4. Cytogenet Cell Genet 1987;46:131–146.
147 Haas M, Rosenberry TL: Quantitative identification of N-terminal amino acids in proteins by radiolabeled reductive methylation of acetylcholinesterase. Anal Biochem 1985;148:74–77.
148 Haas R, Marshall TL, Rosenberry TL: *Drosophila* acetylcholinesterase: Demonstration of a glycoinositol phospholipid anchor and an endogenous proteolytic cleavage. Biochemistry 1988;27:6453–6457.
149 Hada T, Yamawaki M, Moriwaki Y, Tamura S, Yamamoto T, Amuro Y, Nabeshima K, Higashino K: Hypercholinesterasemia with isoenzymic alteration in a family. Clin Chem 1985;31:1997–2000.
150 Hall ZW: Multiple forms of acetylcholinesterase and their distribution in endplate and non-endplate regions of rat diaphragm muscle. J Neurobiol 1973;4:343–361.
151 Hall JG, Pallister PD, Clarren SK, Beckwith JB, Wiglesworth FW, Fraser FC, Cho S, Benke PJ, Reed SD: Congenital hypothalamic hamartoblastoma, hypopituitarism, imperforate anus, and postaxial polydactyly – a new syndrome? I. Clinical, causal and pathogenetic considerations. Am J Med Genet 1980;7:47–74.
152 Hall JC: Genetics of the nervous system in *Drosophila*. Q Rev Biophys 1982;15:3–479.
153 Hall LM, Spierer P: The ace locus of *Drosophila melanogaster:* Structural gene for acetylcholinesterase with an unusual 5′ leader. EMBO J 1986;5:2949–2954.
154 Hedrich HJ, Von Deimling O: Re-evaluation of LGV of the rat and assignment of 12 carboxylesterases to two gene clusters. J Hered 1987;78:92–96.
155 Heijne GV: Signal sequences, the limits of variations. J Mol Biol 1985;184:99–105.
156 Herman RK, Kari CK: Muscle-specific expression of a gene affecting acetylcholinesterase in the nematode *Caenorhabditis elegans*. Cell 1985;40:509–519.
157 Hobbiger F: Reactivation of phosphorylated acetylcholinesterase; in Koelle GB (ed): Cholinesterases and Anticholinesterase Agents. Berlin, Springer, 1963, pp 921–988.
158 Hodgkin WE, Giblett ER, Levine H, Bauer W, Motulsky AG: Complete pseudocholinesterase deficiency: genetic and immunological characterization. J Clin Invest 1965;44:486–493.

159 Hopp TP, Woods KR: Prediction of protein antigenic determinants from amino acid sequences. Proc Natl Acad Sci USA 1981;78:3824–3828.

160 Houamed KM, Bilbe G, Smart TG, Constanti A, Brown DA, Barnard EA, Richards BM: Expression of functional GABA, glycine, and glutamate receptors in *Xenopus* oocytes injected with rat brain mRNA. Nature 1984;310:318–321.

161 Huerre C, Uzan G, Greschik KH, Weil D, Levin M, Hors-Cayla MC, Boue J, Kahn A, Junien C: The structural gene for transferrin TF maps to 3q21-3qter. Ann Genet 1984;40:107–127.

162 Inestrosa NC, Reiness CG, Reichardt LF, Hall ZW: Cellular localization of the molecular forms of acetylcholinesterase in rat pheochromocytoma PC12 cells treated with nerve growth factor. J Neurosci 1981;1:1260–1267.

163 Inestrosa NC, Roberts WL, Marshall TL, Rosenberry TL: Acetylcholinesterase from bovine caudate nucleus is attached to membranes by a novel subunit distinct from those of acetylcholinesterases in other tissues. J Biol Chem 1987;262:4441–4444.

164 Jackson CW: Cholinesterase as a possible marker for early cells of the megakaryocytic series. Blood 1973;42:413–421.

165 Jedrzejczyk J, Silman I, Lai J, Barnard EA: Molecular forms of acetylcholinesterase in synaptic and extrasynaptic regions of avian tonic muscle. Neurosci Lett 1980;46:283–289.

166 Johnson CD, Russell RL: A rapid, simple radiometric assay for cholinesterase, suitable for multiple determinations. Anal Biochem 1975;64:229–238.

167 Johnson CD, Duckett JG, Culotti JG, Herman RK, Meneely PM, Russell RL: An acetylcholinesterase-deficient mutant of the nematode *Caenorhabditis elegans*. J Neurochem 1981;98:261–279.

168 Johnson CD, Russell RL: Multiple molecular forms of acetylcholinesterase in the nematode *Caenorhabditis elegans*. J Neurochem 1983;41:30–36.

169 Johnson JA, Wallace KB: Species-related differences in the inhibition of brain acetylcholinesterase by paroxon and malaoxon. Toxicol Appl Pharmacol 1987;88:234–241.

170 Johnson CD, Rand JB, Herman RK, Stern BD, Russell RL: The acetylcholinesterase genes of *Caenorhabditis elegans:* Identification of a third gene (ace-3) and mosaic mapping of a synthetic lethal phenotype. Neuron 1988;1:165–173.

171 Kalow W, Gunn DR: Some statistical data on atypical cholinesterase of human serum. Ann Hum Genet (Lond) 1959;23:239.

172 Karnovsky MJ, Roots L: A 'direct-coloring' thiocholine method for cholinesterases. J Histochem Cytochem 1964;12:219–221.

173 Karpiak SE, Vilim F, Mahadik SP: Gangliosides accelerate rat neonatal learning and levels of cortical acetylcholinesterase. Dev Neurosci 1983;6:127–135.

174 Kidd KK, Gusella J: Report of the committee on the genetic constitution of chromosomes 3 and 4. Cytogenet Cell Genet 1985;40:107–127.

175 Kim BH, Rosenberry TL: A small hydrophobic domain that localizes human erythrocyte acetylcholinesterase in liposomal membranes is cleaved by papain digestion. Biochemistry 1985;24:3586–3592.

176 Kiraly J, Szentesi I, Ruzicska M, Czeizel A: Chromosome studies in workers producing organophosphate insecticides. Arch Environ Contam Toxicol 1979;8:309–319.

177 Klose R, Gustensohn G: Treatment of alkyl phosphate poisoning with purified serum cholinesterase. Prakl Anasthe 1976;11:1–7.

178 Kodama K, Sikolska H, Bayly R, Bandy-Dafoe P, Wall JR: Use of monoclonal antibodies to investigate a possible role of thyroglobulin in the pathogenesis of Graves' ophthalmopathy. J Clin Endocrinol Metab 1987;59:67–84.

179 Kohl NE: Transposition and amplification of oncogene-related sequences in human neuroblastoma. Cell 1983;35:359–367.

180 Koelle GB, Friedenwald JS: A histochemical method for localizing cholinesterase activity. Proc Soc Exp Biol Med 1949;70:617–622.

181 Koelle GB: Anticholinesterase agents; in Goodman LS, Gilman A (eds): Pharmacological Basis of Therapeutics. ed. 5. New York, Macmillan, 1972, pp 445–466.

182 Koelle WA, Smyrl, EG, Ruch GA, Siddons VE, Koelle GB: Effects of protection of butyrylcholinesterase on regeneration of ganglionic acetylcholinesterase. J Neurochem 1977;28:307–311.

183 Koenig J, Rieger F: Biochemical stability of the AChE molecular forms after cytochemical staining – post-natal localization of the high molecular weight forms of AChE. Dev Neurosci 1981;4:249–257.

184 Kolson DL, Russell RL: New acetylcholinesterase deficient mutants of the nematode *Caenorhabditis elegans*. J Neurogenet 1985;2:69–91.

185 Kolson DL, Russell RL: A novel class of acetylcholinesterase, revealed by mutation in the nematode *Caenorhabditis elegans*. J Neurogenet 1985;2:93–110.

186 Kostovic I, Goldman-Rakic PS: Transient cholinesterase staining in the mediodorsal nucleus of the thalamus and its connections in the developing human and monkey brain. J Comp Neurol 1983;219:431–447.

187 Krieg PA, Melton DA: Functional messenger RNAs are produced by SP6 in vitro transcription of cloned cDNAs. Nucleic Acids Res 1984;12:7057–7070.

188 Kristt DA: Acetylcholinesterase in the ventrobasal thalamus: Transience and patterning during ontogenesis. Neuroscience 1983;10:923–939.

189 Kusano K, Miledi R, Stinnakre J: Acetylcholine receptors in the oocyte membrane. Nature 1977;270:739–741.

190 Kyte J, Doolittle RF: A simple model for displaying the hydropathic character of sequence segments. J Mol Biol 1982;157:105–132.

191 Lapidot-Lifson Y, Prody CA, Ginzberg D, Meytes D, Zakut H, Soreq H: Co-amplification of human acetylcholinesterase and butyrylcholinesterase genes in blood cells. Correlation with various leukemias and abnormal megakaryocytopoiesis. Proc Natl Acad Sci USA 1989;86:4715–4719.

192 Lappin R, Lee I, Lieberburg IM: Generation of subunit specific antibody probes for *Torpedo* acetylcholinesterase: Cross-species reactivity and use of cell-free translations. J Neurobiol 1987;18:75–99.

193 Laskowski MB, Detbarn WD: An electrophysiological analysis of the effect of paraoxon at the neuromusclar junction. J Pharmacol Exp Ther 1979;210:269.

194 Lathe R: Synthetic oligonucleotide probes deduced from amino acid sequence data: Theoretical and practical considerations. J Mol Biol 1985;183:1–12.

195 Layer PG: Comparative localization of acetylcholinesterase and pseudocholinesterase during morphogenesis of the chicken brain. Proc Natl Acad Sci USA 1983;80:6413–6417.

196 Layer PG, Sporns O: Spatiotemporal relationship of embryonic cholinesterases with cell proliferation in chicken brain and eye. Proc Natl Acad Sci USA 1987;84:284–288.

197 Layer PG, Alber R, Sporns O: Quantitative development and molecular forms of acetyl- and butyrylcholinesterase during morphogenesis and synaptogenesis of chick brain and retina. J Neurochem 1987;49:175–182.

198 Layer PG, Alber R, Rathjen FG: Sequential activation of butyrylcholinesterase in rostral half somites and acetylcholinesterase in motorneurons and myotomes preceding growth of motor axons. Development 1988;102:387–396.

199 Layer PG, Rommel S, Bulthoff H, Hengstenberg R: Independent spatial waves of biochemical differentiation along the surface of chicken brain as revealed by the sequential expression of acetylcholinesterase. Cell Tissue Res 1988;251:587–595.

200 Lee TC, Tanaka N, Lamb PW, Gilmer TM, Barrett JC: Arsenic induces gene amplification in vivo. Science 1988;241:79–81.

201 Levanon D, Lieman-Hurwitz J, Dafni N, Wigderson M, Bernstein Y, Laver-Rudich Z, Danciger E, Stein O, Groner Y: Architecture and anatomy of the chromosomal locus in human chromosome 21 encoding the CuZn-superoxide dismutase. EMBO J 1985;4:77–84.

202 Levey AI, Wainer BH, Mufson EJ, Mesulam MM: Colocalization of acetylcholinesterase and choline acetyltransferase in rat cerebrum. Neuroscience 1983;9:9–22.

203 Libermann TA, Razon N, Bartal AD, Yarden T, Schlessinger J, Soreq H: Expression of epidermal growth factor receptors in human brain tumors. Cancer Res 1984;44:753–760.

204 Libermann TA, Nusbaum HR, Razon N, Kris R, Lax I, Soreq H, Whittle N, Waterfield MD, Ullrich A, Schlessinger J: Amplification, enhanced expression and possible rearrangement of EGF receptor gene in primary human brain tumors of glial origin. Nature 1985;313:144–147.

205 Lindstrom JM, Seybold ME, Lennon BA, Whittinham S, Duane DD: Antibody to acetylcholine receptor in myasthenia gravis. Prevalence, clinical correlates, and diagnostic value. Neurology 1976;26:1054.

206 Livneh A, Sarova-Pinhas I, Pras M, Wagner K, Zakut H, Soreq H: Antibodies against acetylcholinesterase and low levels of cholinesterases in a patient with an atypical neuromuscular disorder. Clin Immunol Immunopathol 1988;48:119–131.

207 Lockridge O: Amino acid composition and sequence of human serum cholinesterase: A progress report; in Barnard EA, Sket D (eds): Cholinesterase Fundamental and Applied Aspects. New York, de Gruyter, 1984, pp 5–12.

208 Lockridge O, La Du BN: Amino acid sequence of the active site of human serum cholinesterase from usual, atypical and atypical-silent genotypes. Biochem Genet 1986;24:485–498.

209 Lockridge O, Bartels CG, Vaughan TA, Wong CK, Norton SE, Johnson LL: Complete amino acid sequence of human serum cholinesterase. J Biol Chem 1987;262:549–557.

210 Loomis TA: Distribution and excretion of pyridine-2-aldoxime methiodide (PAM): Atropine and PAM in sarin poisoning. Toxicol Appl Pharmacol 1963;5:489–499.

211 Lovrien EW, Magenis RE, Rivas ML, Lamvik N, Rowe S, Wood J, Hemmerling J: Serum cholinesterase E2 linkage analysis: Possible evidence for localization to chromosome 16. Cytogenet Cell Genet 1978;22:324–326.

212 Ludgate M, Owada C, Pope R, Taylor P, Vassart G: Cross-reactivity between antibodies to human thyroglobulin and *Torpedo* acetylcholinesterase in patients with Graves' ophthalmopathy (abstract). J Endocr Invest 1986;9(suppl 3):89.

213 Ludgate M, Swillens S, Mercken L, Vassart G: Homology between thyroglobulin and acetylcholinesterase: An explanation for pathogenesis of Graves' ophthalmopathy. Lancet 1986;ii:219.

214 Ludgate M, Dong Q, Dreyfus PA, Zakut H, Taylor P, Vessart G, Soreq H: Definition, at the molecular level, of a thyroglobulin-acetylcholinesterase shared epitope: Study of its pathophysiological significance in patients with Graves' ophthalmopathy. Autoimmunity (in press, 1990).

215 Lyles JM, Silman I, Di Giamberardino L, Couraud JY, Barnard EA: Comparison of the molecular forms of the cholinesterases in tissues of normal and dystrophic chickens. J Neurochem 1982;38:1007–1021.

216 MacPhee-Quigley K, Taylor P, Taylor S: Primary structures of the catalytic subunits from two molecular forms of acetylcholinesterase: A comparison of NH2-terminal and active center sequences. J Biol Chem 1985;260:12185–12189.

217 MacPhee-Quigley K, Vedvick TS, Taylor P, Taylor SS: Profile of the disulfide bonds in acetylcholinesterase. J Biol Chem 1986;261:13565–13570.
218 Malinger G: Expression of human cholinesterase genes in normal and malignant ovarian tissues; Basic Sciences Thesis, The Weizmann Institute, Rehovot, Israel, 1987.
219 Malinger G, Zakut H, Soreq H: Cholinoceptive properties of human primordial, pre-antral and antral oocytes: in situ hybridization and biochemical evidence for expression of CHE genes. J Molec Neurosci 1989;1:77–84.
220 Mane SD, Tompkins L, Richmond RC: Male esterase 6 catalyzes the synthesis of a sex pheromone in *Drosophila melanogaster* females. Science 1983;222:419–421.
221 Marquis JK, Fishman EB: Presynaptic acetylcholinesterase. Trends Pharm Sci 1985(Oct):387–388.
222 Marsh D, Grassi J, Vigny M, Massoulie J: An immunological study of rat acetylcholinesterase: Comparison with acetylcholinesterase from other vertebrates. J Neurochem 1984;43:204–213.
223 Martin FH, Castro M, Aboul-ela M, Tinoco I Jr: Base pairing involving deoxyinosine: Implications for probe design. Nucleic Acids Res 1985;13:8927–8938.
224 Martson L, Voronina V: Experimental study of the effect of a series of phosphoroorganic pesticides (Dipterex and Imidan) on embryogenesis. Environ Health Perpect 1976;13:121–125.
225 Massoulie J, Bon S: The molecular forms of cholinesterase and acetylcholinesterase in vertebrates. Annu Rev Neurosci 1982;5:57–106.
226 Mattei MG, Philip N, Passage E, Moisan JP, Mandel JL, Mattei JE: DNA probe localization of 18p11,13 band by in situ hybridization and identification of a small supernumerary chromosome. Hum Genet 1985;69:268–271.
227 McGuire MC, Nogueira CP, Bartels CF, Lightstone H, Hajra A, Van der Spek ASL, Lockridge O, La Du BN: Identification of the structural mutation responsible for the dibucaine-resistant (atypical) variant form of human serum cholinesterases. Proc Natl Acad Sci USA 1989;86:953–957.
228 McTiernan C, Adkins S, Chantonnet A, Vaughan TA, Bartels CF, Kott M, Rosenberry TL, La Du BN: Brain cDNA clone for human cholinesterase. Proc Natl Acad Sci USA 1987;84:6682–6686.
229 Meflah K, Bernard S, Massoulie J: Interactions with lectins indicate differences in the carbohydrate composition of the membrane-bound enzymes acetylcholinesterase and 5′-nucleotidase in different cell types. Biochemistry 1984;66:59–69.
230 Melchior WB, Hippel HV: Alteration of the relative stability of dA-dT and dG-dC base pairs in DNA. Proc Natl Acad Sci USA 1972;70:298–302.
231 Mengitsu M, Laryea E, Miller A, Wall JR: Clinical significance of a new autoantibody against a human eye muscle soluble antigen, detected by immunofluorescence. Clin Exp Immunol 1986;65:19–27.
232 Merken L, Simons MJ, Swillens S, Massaer M, Vassart G: Primary structure of bovine thyroglobulin deduced from the sequence of its 8,431-base complementary DNA. Nature 1985;316:647–651.
233 Mintz KP, Brimijoin S: Monoclonal antibodies to rabbit brain acetylcholinesterase: Selective enzyme inhibition, differential affinity for enzyme forms, and cross-reactivity with other mammalian cholinesterases. J Neurochem 1985;45:284–292.
234 Mishina M, Kurosaki T, Tobimatsu T, Morimoto Y, Noda M, Yamamoto T, Terao M, Lindstrom J: Expression of functional acetylcholine receptor from cloned cDNAs. Nature 1984;307:604–608.
235 Mollgard K, Dziegielewska KM, Saunders NR, Zakut H, Soreq H: Synthesis and localization of plasma proteins within specific cell types in the developing fetal human

brain: correlation of mRNA translation with immunocytochemistry. Dev Biol 1988;128:207–221.

236 Morel N, Dreyfus PA: Association of acetylcholinesterase with the external surface of the presynaptic plasma membrane in *Torpedo* electric organ. Neurochem Int 1982;4:283–288.

237 Moutschen-Dahmen J, Moutschen-Dahmen M, Degraeve N: Mutagenicity, Carcinogenicity and Teratogenicity of Industrial Pollutants. New York, Plenum Press, 1980, pp 127–203.

238 Mulder DW, Lambert EH, Eaton LU: Myasthenic syndrome in patients with amyotrophic lateral sclerosis. Neurology 1959;9:627.

239 Mullin BR, Levinson RE, Friedman A, Henson DE, Winand RJ, Kohn LD: Delayed hypersensitivity in Graves' disease and exophthalmos: identification of thyroglobulin antibodies in normal human orbital muscle. Endocrinology 1977;100:351–366.

240 Musset F, Frobert Y, Grassi J, Vigny M, Boulla G, Bon S, Massoulie J: Monoclonal antibodies against acetylcholinesterase from electric organs of *Electorphorus* and *Torpedo*. Biochemistry 1987;69:147–156.

241 Myers M, Richmond RC, Oakeshott JG: On the origin of esterases. Mol Biol Evol 1988;5:785–796.

242 Nagoshi RN, Gelbart WN: Genetics of the ace locus in *Drosophila melanogaster*. Genetics 1987;117:487–502.

243 Namba T, Nolte CT, Jackrel J, Grob D: Poisoning due to organophosphate insecticides: acute and chronic manifestations. Am J Med 1971;50:475–492.

244 Neurath H: Evolution of proteolytic enzymes. Science 1984;224:350–357.

245 Nicholas A, Vienne M, Van Den Berghe H: Induction of sister chromatid exchanges in cultured human cells by an organophosphorous insecticide: malathion. Mutat Res 1979;67:167–172.

246 Nicolet M, Pincon-Raymond M, Rieger F: Globular and asymmetric acetylcholinesterase in frog muscle basal lamina sheaths. J Cell Biol 1986;102:1–7.

247 Noda M, Takahashi H, Tanabe T, Toyasato M, Kikyotani S, Furutani Y, Hirosi T, Takashima H, Inayama S, Niyata T, Numa S: Structural homology of *Torpedo californica* acetylcholine receptor subunits. Nature 1983;302:528–531.

248 Ogi D, Hamada A: Case reports on fetal deaths and extremities malformations probably related to insecticide poisoning. J Jpn Obstet Gynecol Soc 1965;17:569.

249 Ohno S: Evolution by Gene Duplication. New York, Springer, 1970.

250 Olney JW, de Gubareff T, Labruyere J: Seizure-related brain damage induced by cholinergic agents. Nature 1983;301:520–522.

251 Opresko LK, Karpf RA: Specific proteolysis regulates fusion between endocytotic compartments in *Xenopus* oocytes. Cell 1987;51:557–568.

252 Ord GM, Thompson RHS: Pseudocholinesterase activity in the central nervous system. Biochem J 1952;51:245–251.

253 Oron Y, Dascal N, Nadler E, Lupu M: Inositol 1,4,5-triphosphate mimics muscarinic response in *Xenopus* oocytes. Nature 1985;313:141–143.

254 Ott P, Lustig A, Brodbeck U, Rosenbusch JP: Acetylcholinesterase from human erythrocyte membranes: Dimers as functional units. FEBS Lett 1982;138:187–189.

255 Ott P, Ariano BH, Binggeli Y, Brodbeck U: A monomeric form of human erythrocyte membrane acetylcholinesterase. Biochem Biophys Acta 1983;729:193–199.

256 Ott P: Membrane acetylcholinesterases: Purification, molecular properties and interactions with amphiphilic environments. Biochem Biophys Acta 1985;822:375–392.

257 Palecek J, Habrova V, Nedvidek J, Romanovsky A: Dynamics of tubulin structures in *Xenopus laevis* oogenesis. J Embryol Exp Morphol 1985;87:75–86.

258 Paulus JP, Maigen J, Keyhani E: Mouse megakaryocytes secrete acetylcholinesterase. Blood 1981;58:1100–1106.

259 Pelham HRB, Jackson RJ: An efficient mRNA-dependent translation system from reticulocyte lysates. Eur J Biochem 1976;67:247–256.

260 Perry EK: The cholinergic hypothesis – ten years on. Br Med Bull 1986;42:63–69.

261 Phillips TM, Manz HJ, Smith FA, Jaffe HA, Cohan SL: The detection of anti-cholinesterase antibodies in myasthenia gravis. Ann NY Acad Sci 1981;337:360.

262 Pintada T, Ferro MT, San Roman C, Mayayo M, Larana JG: Clinical correlations of the 3q21;q26 cytogenetic anomaly. A leukemic or myelodysplastic syndrome with preserved or increased platelet production and lack of response to cytotoxic drug therapy. Cancer 1985;55:535–541.

263 Prody C, Zevin-Sonkin D, Gnatt A, Koch R, Zisling R, Goldberg O, Soreq H: Use of synthetic oligodeoxynucleotide probes for the isolation of a human cholinesterase cDNA clone. J Neurosci Res 1986;16:25–35.

264 Prody C, Zevin-Sonkin D, Gnatt A, Goldberg O, Soreq H: Isolation and characterization of full length cDNA clones for cholinesterase from fetal human tissues. Proc Natl Acad Sci USA 1987;84:3555–3559.

265 Prody C, Dreyfus PA, Zamir R, Zakut H, Soreq H: De novo amplification within a 'silent' human cholinesterase gene in a family subjected to prolonged exposure to organophosphorous insecticide. Proc Natl Acad Sci USA 1989;86:860–864.

266 Queen CL, Korn LJ: Computer analysis of nucleic acids and proteins. Methods Enzymol 1980;65:595–609.

267 Rabin M, Fries R, Singer D, Ruddle FH: NARS transforming gene maps to region p11–p13 on chromosome No. 1 by hybridization. Cytogenet Cell Genet 1985;39:206–209.

268 Radic Z, Reiner E, Simeon V: Binding sites on acetylcholinesterase for reversible ligands and phosphorylating agents: A theoretical model tested on haloxon and phosphostigmine. Biochem Pharmacol 1984;33:671–677.

269 Rakonczay Z, Brimijoin S: Monoclonal antibodies to rat brain acetylcholinesterase: Comparative affinity for soluble and membrane-associated enzyme and for enzyme from different vertebrate species. J Neurochem 1986;46:280–287.

270 Rakonczay Z, Brimijoin S: Biochemistry and pathophysiology of the molecular forms of cholinesterases; in Harris JR (ed): Subcellular Biochemistry 12. New York, Plenum Press, 1988, pp 335–378.

271 Rakonczay Z: Cholinesterase and its molecular forms in pathological states. Prog Neurobiol 1988;31:311–330.

272 Rama Sastry BV, Sadavongvivad C: Cholinergic systems in non-nervous tissues. Pharmacol Rev 1979;30:65–132.

273 Ratner D, Oren B, Vigder K: Chronic dietary anticholinesterase poisoning. Int J Med Sci 1983;19:810–814.

274 Razon N, Soreq H, Roth E, Bartal A, Silman I: Characterization of activities and forms of cholinesterases in human primary brain tumors. Exp Neurol 1984;84:681–695.

275 Rieger F, Faivre-Bauman A, Benda P, Vigny M: Molecular forms of acetylcholinesterase: Their de novo synthesis in mouse neuroblastoma cells. J Neurochem 1976;2:1059–1063.

276 Rieger F, Chetelat R, Nicolet M, Kamal L, Poullet M: Presence of tailed, asymmetric forms of acetylcholinesterase in the central nervous system of vertebrates. FEBS Lett 1980;121:169–174.

277 Roberts WL, Kim BH, Rosenberry TL: Differences in the glycolipid membrane anchors of bovine and human erythrocyte acetylcholinesterases. Proc Natl Acad Sci USA 1987;84:7817–7821.

278 Rogers J: Exon shuffling and intron insertion in serine protease genes. Nature 1985;315:458–459.

279 Rosenberry TL: Acetylcholinesterase. Adv Enzymol 1975;43:103–218.

280 Rosenberry TL, Richardson JM: Structure of 18S and 14S acetylcholinesterase. Identification of collagen-like subunits that are linked by disulfide bonds to catalytic subunits. Biochemistry 1977;16:3550–3558.

281 Rosenberry TL, Scoggin DM: Human erythrocyte acetylcholinesterase is an amphipathic protein whose short membrane-binding domain is removed by papain digestion. J Biol Chem 1984;250:5643–5652.

282 Rosenfeld MG, Amara SG, Ross BA, Ong ES, Evans RM: Altered expression of the calcitonic gene associated with RNA polymorphism. Nature 1981;290:63–65.

283 Roskoski RJR: Choline acetyltransferase and acetylcholinesterase: evidence for essential histidine residues. Biochemistry 1974;13:5141–5144.

284 Rotundo RL: Asymmetric acetylcholinesterase is assembled in the Golgi apparatus. Proc Natl Acad Sci USA 1984;81:479–483.

285 Rotundo RL: Biogenesis and regulation of acetylcholinesterases; in The Vertebrate Neuromuscular Junction. New York, Liss, 1987, pp 247–284.

286 Rotundo RL: Carbonetto ST: Neurons segregate clusters of membrane-bound acetylcholinesterase along their neurites. Proc Natl Acad Sci USA 1987;84:2063–2067.

287 Rotundo RL, Gomez AM, Fernandez-Valle C, Randall WR: Allelic variants of acetylcholinesterase: Genetic evidence that all acetylcholinesterase forms in avian nerves and muscles are encoded by a single gene. Proc Natl Acad Sci USA 1988;85:7805–7809.

288 Ruberg M, Rieger F, Villageois A, Bunnet AM, Agid Y: Acetylcholinesterase and butyrylcholinesterase in frontal cortex and cerebrospinal fluid of demented and non-demented patients with Parkinson's disease. Brain Res 1986;362:83–91.

289 Slamon DJ, Clark GM, Wong SG, Levin WJ, Ullrich A, McGuire WL: Human breast cancer: correlation of relapse and survival with amplification of the HER-2/Neu oncogene. Science 1987;235:177–182.

290 Salpeter M: Electron microscope radioautography as a quantitative tool in enzyme cytochemistry. I. The distribution of acetylcholinesterase at motor endplates of a vertebrate twitch muscle. J Cell Biol 1967;32:339–389.

291 Sanger G, Nicklen S, Coulson AR: DNA sequencing with chain-terminating inhibitors. Proc Natl Acad Sci USA 1977;74:5463–5468.

292 Sanyal RK, Khanna SK: Action of cholinergic drugs on motility of spermatozoa. Fertil Steril 1971;22:356–359.

293 Schimke RT: Gene amplification in cultured animal cells. Cell 1984;37:705–713.

294 Schumacher M, Camp S, Maulet Y, Newton M, MacPhee-Quigley K, Taylor SS, Freidmann T, Taylor P: Primary structure of *Torpedo californica* acetylcholinesterase deduced from its cDNA sequence. Nature 1986;319:407–409.

295 Schwab M, Varmus HE, Bishop JM: Human N-myc gene contributes to neoplastic transformation of mammalian cells in culture. Nature 1985;316:160–163.

296 Scoto KW, Biedler JL, Melera PW: Amplification and expression of genes associated with multidrug resistance in mammalian cells. Science 1986;232:751–755.

297 Seidman S, Soreq H: Co-injection of *Xenopus* oocytes with cDNA-produced and native mRNAs: a molecular biological approach to the tissue specific processing of human cholinesterases. Int Rev Neurobiol, in press.

298 Seiler J: Inhibition of testicular DNA synthesis by chemical mutagens and carcinogens. Mutat Res 1977;46:305–310.

299 Sikorska H, Wall JR: Failure to detect eye muscle membrane specific autoantibodies in Graves' ophthalmopathy. Br Med J 1985;291:604–607.

300 Sikorav JL, Krejci E, Massoulie J: cDNA sequences of *Torpedo marmorata* acetylcholinesterase: Primary structure of the precursor of a catalytic subunit; Existence of multiple 5'-untranslated regions. EMBO J 1987;6:1865–1873.
301 Sikorav JL, Duval N, Anselmet A, Bon S, Krejici E, Legay C, Osterlund M, Reimund B, Massoulie J: Complex alternative splicing of acetylcholinesterase transcripts in *Torpedo* electric organ; primary structure of the precursor of the glycolipid-anchored dimeric form. EMBO J 1988;7:2983–2993.
302 Silman I, Dudai Y: Molecular structure and catalytic activity of membrane-bound acetylcholinesterase from electric organ tissue of the electric eel; in Reiner E (ed): Cholinesterases. Amsterdam, North-Holland, 1975, pp 181–200.
303 Silman I, DiGiambernardino L, Lyles J, Couraud JY, Barnard EA: Parallel regulation of acetylcholinesterase and pseudocholinesterase in normal, denervated and dystrophic chicken skeletal muscle. Nature 1979;280:160–161.
304 Silman I, Futerman AH: Modes of attachment of acetylcholinesterase to the surface membrane. Eur J Biochem 1987;170:11–22.
305 Silver A: The Biology of Cholinesterases. Amsterdam, North-Holland, 1974.
306 Simmers RN, Stupans I, Sutherland GR: Localization of the human haptoglobin genes distal to the fragile site at 16q22 using in situ hybridization. Cytogenet Cell Genet 1986;41:38–41.
307 Simpson NE: Factors influencing cholinesterase activity in a Brazilian population. Am J Hum Gen 1966;18(No. 3).
308 Smith AD, Cuello AC: Alzheimer's disease and acetylcholinesterase-containing neurons. Lancet 1984;i:513.
309 Sorensen K, Getinetta R, Brodbeck U: An amphiphile-dependent form of human brain caudate nucleus acetylcholinesterase: Purification and properties. J Neurochem 1982;39:1050–1060.
310 Sorensen K, Brodbeck U, Rasmussen AG, Norgaard-Pedersen B: An inhibitory monoclonal antibody to human acetylcholinesterases. Biochim Biophys Acta 1987;912:56–62.
311 Soreq H, Parvari R, Silman I: Biosynthesis and secretion of catalytically active acetylcholinesterase in *Xenopus* oocytes microinjected with mRNA from *Torpedo* electric organ. Proc Natl Acad Sci USA 1982;79:830–834.
312 Soreq H, Zevin-Sonkin D, Razon N: Expression of cholinesterase gene(s) in human brain tissues: Translational evidence for multiple mRNA species. EMBO J 1984;3:1371–1375.
313 Soreq H: The biosynthesis of biologically active proteins in RNA microinjected *Xenopus* oocytes. CRC Crit Rev Biochem 1985;18:199–238.
314 Soreq H, Dziegielewska KM, Zevin-Sonkin D, Zakut H: The use of mRNA translation in vitro and in ovo followed by crossed immunoelectrophoretic autoradiography to study the biosynthesis of human cholinesterases. Cell Molec Neurobiol 1986;6:227–237.
315 Soreq H, Gnatt A: Molecular biological search for human genes encoding cholinesterases. Mol Neurobiol 1987;1:47–80.
316 Soreq H, Malinger G, Zakut H: Expression of cholinesterase genes in human oocytes revealed by in situ hybridization. Hum Reprod 1987;2:689–693.
317 Soreq H, Zamir R, Zevin-Sonkin D, Zakut H: Human cholinesterase genes localized by hybridization to chromosomes 3 and 16. Hum Genet 1987;77:325–328.
318 Soreq H, Prody CA: Sequence similarities between human acetylcholinesterase and related proteins: Putative implications for therapy of anticholinesterase intoxication; in Golombek A, Rein R (eds): Computer-Assisted Modeling of Receptor-Ligand Interactions: Theoretic Aspects and Application to Drug Design. New York, Liss, 1989, pp 347–359.

319 Soreq H, Seidman S, Dreyfus PA, Sonkin D, Zakut H: Expression and tissue-specific assembly of cholinesterase in SP6mRNA-injected oocytes. J Biol Chem 1989;264:10608–10613.

320 Soreq H, Zakut H: Amplification of acetylcholinesterase and butyrylcholinesterase genes in normal and tumor tissues: Putative relationship to organophosphorous poisoning. Pharm Res 1990;7:1–7.

321 Sparkes RS, Field LL, Sparkes MC, Crist M, Spence MA, Janes K, Garry PJ: Genetic linkage studies of transferrin, pseudocholinesterase, and chromosome 1 loci. Hum Hered 1984;34:96–100.

322 Speiss M, Lodish HF: An internal signal sequence: the asialoglycoprotein receptor anchor. Cell 1986;44:177–185.

323 Spradling A: Structure and Function of Eukaryotic Chromosomes. Berlin, Springer, 1987, vol 14, pp 200–212.

324 Stanley KK, Luzio JP: Construction of a new family of high efficiency bacterial expression vectors: Identification of cDNA clones coding for human liver proteins. EMBO J 1984;3:1429–1434.

325 Stark GR: Gene amplification in drug-resistant cells and in tumors. Cancer Surv 1986;5:1–23.

326 Sytkowski AJ, Vogel Z, Nirenberg MW: Development of acetylcholine receptor clusters on cultured muscle cells. Proc Natl Acad Sci USA 1973;70:270–274.

327 Sweet DL, Golomb HM, Rowley JD, Vardiman JM: Acute myelogenous leukemia and thrombocythemia associated with an abnormality of chromosome No. 3. Cancer Genet Cytogenet 1979;1:33–37.

328 Swillens S, Ludgate M, Mercken L, Dumont JE, Vessart G: Analysis of sequence and structure homologies between thyroglobulin and acetylcholinesterase: possible functional and clinical significance. Biochem Biophys Res Commun 1986;137:142–148.

329 Szeinberg A, Pipano S, Assa M, Medalie JH, Neufeld HN: High frequency of atypical pseudocholinesterase gene among Iraqi and Iranian Jews. Clin Genet 1972;3:123–127.

330 Szelenyi JG, Bartha E, Hollan SR: Acetylcholinesterase activity of lymphocytes: An enzyme characteristic of T-cells. Br J Haematol 1982;50:241–245.

331 Tao TW, Cheng PJ, Phan H, Leu SL, Kriss JP: Monoclonal anti-thyroglobulin antibodies derived from immunizations of mice with human eye muscle and thyroid membranes. J Clin Endocrinol Metab 1986;63:577–582.

332 Taub R, Kelly J, Latt S, Lenoir GM, Tantravahi TV, Tu Z, Leder P: A novel alteration in the structure of an activated c-myc gene in a variant t(2;8) Burkitt lymphoma. Cell 1984;37:511–520.

333 Taylor P, Schumacher M, MacPhee-Quigley K, Friedmann T, Taylor S: The structure of acetylcholinesterase: Relationship to its function and cellular disposition. Trends Neurosci 1987;10:93–95.

334 Thesleff S: The mode of neuromuscular block caused by acetylcholine, nicotine, decamethonium and succinylcholine. Acta Physiol Scand 1955;34:218.

335 Tiedt TN, Albuquerque EX, Hudson CS, Rash JE: Neostigmine-induced alterations at the mammalian neuromuscular junction. I. Muscle construction and electrophysiology. J Pharmacol Exp Ther 1978;205:326.

336 Toutant JP, Massoulie J, Bon S: Polymorphism of pseudocholinesterase in *Torpedo marmorata* tissues: Comparative study of the catalytic and molecular properties of this enzyme with acetylcholinesterase. J Neurochem 1987;44:580–592.

337 Toutant JP, Massoulie J: Acetylcholinesterase; in Kenny, Turner (eds): Mammalian Ectoenzymes. Amsterdam, Elsevier, 1985, pp 289–328.

338 Toutant JP, Massoulie J: Cholinesterase II – Tissue and cellular distribution of the molecular forms and their physiological regulations; in Whittaker VP (ed): The Cholinergic Synapse. Handbook of Experimental Pharmacology. Berlin, Springer, 1988.
339 Trinh Van Bao, Szabo I, Ruzicska P, Czeizel A: Chromosome aberrations in patients suffering acute organic phosphate insecticide intoxication. Human Genet 1974;24:33–57.
340 Tsim KWK, Randall WR, Barnard EA: Monoclonal antibodies specific for the different subunits of asymmetric acetylcholinesterase from chick muscle. J Neurochem 1988;51:95–104.
341 UN Security Council: Report of specialists appointed by the Secretary General. Paper S/16433;1984.
342 Vigny M, Gisiger V, Massoulie J: 'Nonspecific' cholinesterase and acetylcholinesterase in rat tissues: Molecular forms, structural and catalytic properties, and significance of the two enzyme systems. Proc Natl Acad Sci USA 1978;75:2588–2592.
343 Wallace BG, Nitkin RM, Reist NE, Fallon JR, Moayeri NN, McMahan UJ: Aggregates of acetylcholinesterase induced by acetylcholine-receptors aggregating factor. Nature 1985;315:574–577.
344 Wallace BG: Aggregating factor from *Torpedo* electric organ induces patches containing acetylcholine receptors, acetylcholinesterase, and butyrylcholinesterase on cultured myotubes. J Cell Biol 1986;102:783–794.
345 Wecker L, Kiauta T, Dettbarn WD: Relationship between acetylcholinesterase inhibition and the development of a myopathy. J Pharmacol Exp Ther 1978;206:97–104.
346 Weeks DL, Melton DA: A maternal mRNA localized in the vegetal hemisphere in *Xenopus* eggs codes for a growth factor related to TGF-beta. Cell 1987;51:861–867.
347 Whittaker M: Plasma cholinesterase and the anaesthetist. Anaesthesia 1980;35:174–197.
348 Whittaker M: Cholinesterase: Monographs in Human Genetics. Basel, Karger, 1986, vol 11.
349 Wischnitzer S: The ultrastructure of the cytoplasm of the developing amphibian egg; in Abercrombie M, Brachet J (eds): Advances in Morphogenesis. New York, Academic Press, 1966, vol 5, pp 131–179.
350 Wollemann M, Zoltan L: Cholinesterase activity of cerebral tumorous cysts. Arch Neurol Psychiatry 1962;6:161–167.
351 Wong AJ, Bigner SH, Bigner DD, Kinzler KW, Hamilton AR, Vogelstein B: Increased expression of the epidermal growth factor receptor gene in malignant gliomas is invariably associated with gene amplification. Proc Natl Acad Sci USA 1987;84:6899–6903.
352 Woodland HR, Wilt FH: The functional stability of sea urchin histone mRNA injected into oocytes of *Xenopus laevis*. Dev Biol 1980;75:199–204.
353 Wyrobek A, Bruce W: Chemical induction of sperm abnormalities in mice. Proc Natl Acad Sci USA 1975;72:4425–4429.
354 Yates CM, Simpson J, Maloney AFJ, Gordon A, Reid AH: Alzheimer-like cholinergic deficiency in Down's syndrome. Lancet 1980;ii:979–980.
355 Yoder J, Watson M, Benson W: Lymphocyte chromosome analysis of agricultural workers during extensive occupational exposure to pesticides. Mutat Res 1973;21:335–340.
356 Young KM, Weiss L: Megakaryocytopoiesis: Incorporation of tritiated thymidine by small acetylcholinesterase-positive cells in murine bone marrow during antibody-induced thrombocytopenia. Blood 1987;69:290–295.
357 Zakut H, Matzkel A, Schejter E, Avni A, Soreq H: Polymorphism of acetylcholinesterase in discrete regions of the developing human fetal brain. J Neurochem 1985;45:382–389.

358 Zakut H, Even L, Birkenfeld S, Malinger G, Zisling R, Soreq H: Modified properties of serum cholinesterases in primary carcinomas. Cancer 1988;61:727–737.
359 Zakut H, Zamir R, Sindell L, Soreq H: Gene mapping on chorionic villi chromosomes by hybridization in situ: Refinement of cholinesterase cDNA binding sites to chromosome 3q21, 3q26 and 16q21. Hum Reprod 1989;4:941–946.
360 Zon G: Oligonucleotide analogues as potential chemotherapeutic agents. Pharm Res 1988;5:539–549.

Subject Index

12-6-78 associated Industries 6 35

To make the puppets in this book, all you need are gloves, mittens, socks, scissors, needle and thread, glue, string, and buttons. For some puppets you will also need yarn, tissues, cloth scraps, paper, pencil, a cotton ball, and a pipe cleaner.

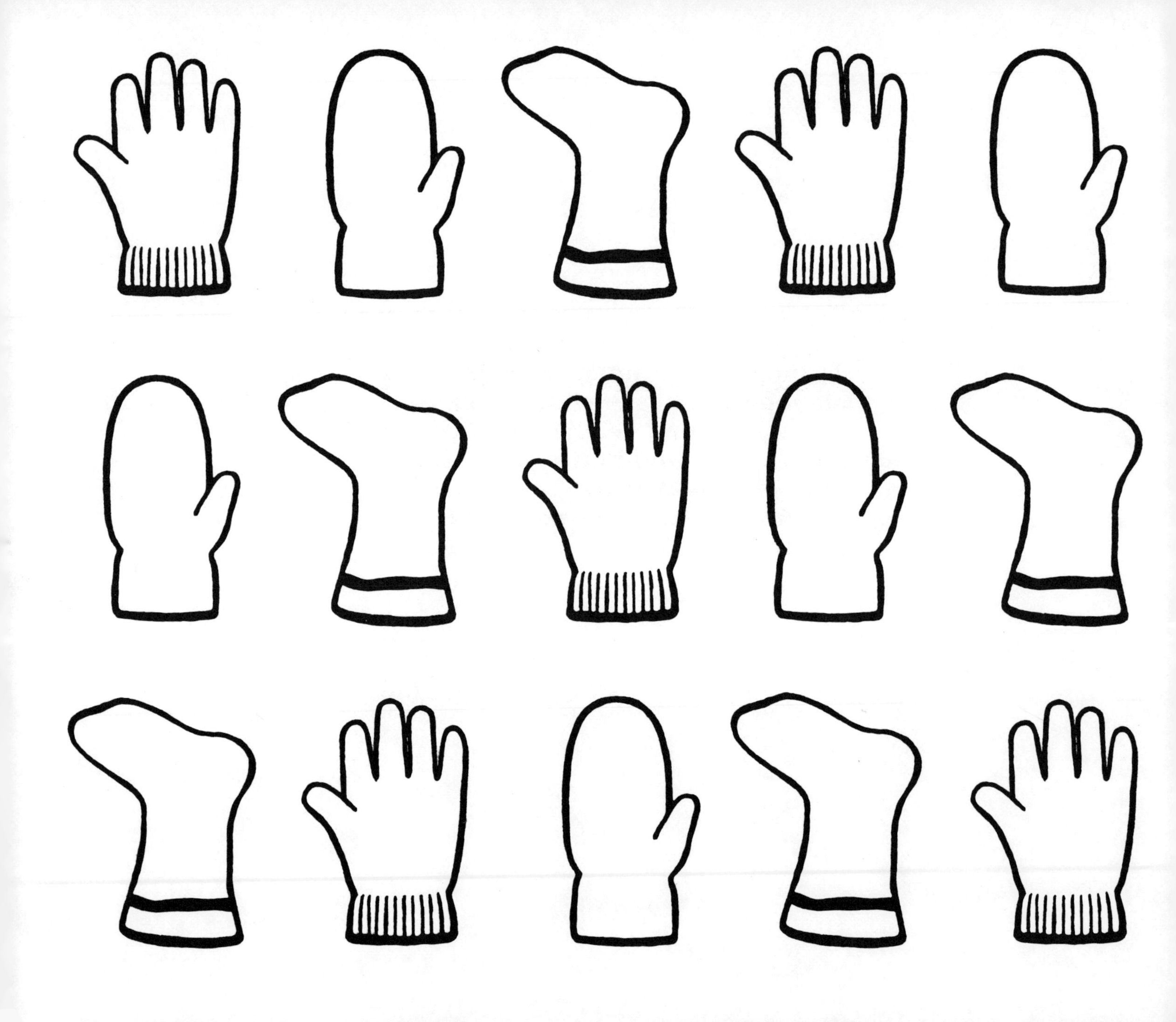

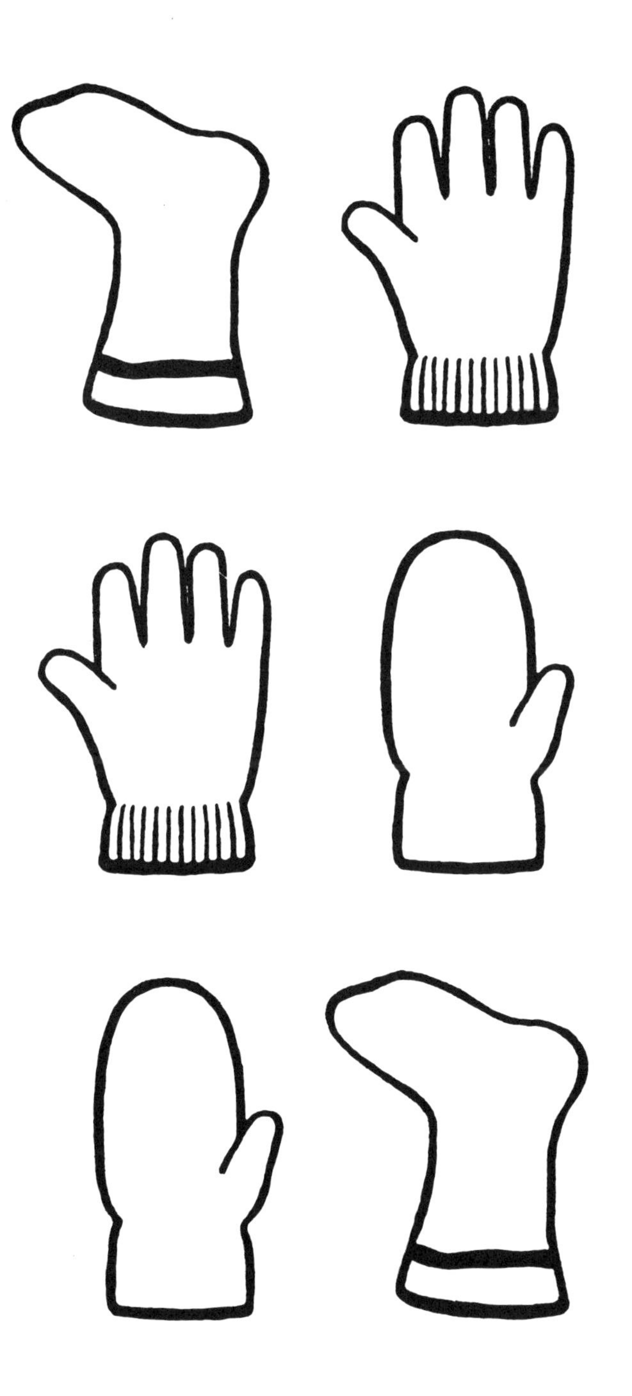

glove, mitten, and sock puppets

by Frieda Gates

WALKER AND COMPANY, NEW YORK

CONTENTS

Published in the United States of America in 1978 by the Walker Publishing Company, Inc.

Published simultaneously in Canada by Beaverbooks, Ltd., Pickerin Ontario

ISBN: 0-8027-6326-X

Reinf. ISBN: 0-8027-6327-8

Library of Congress Catalog Card Number 78-3068

Printed in the United States of America

FINGER PUPPETS

Cut the fingers off an old glove.

Sew on buttons or beads for eyes and noses.

Cut out felt ears, mouths, and tails. Use yarn for hair.

Sew hair, ears, mouths, and tails on finger puppets.

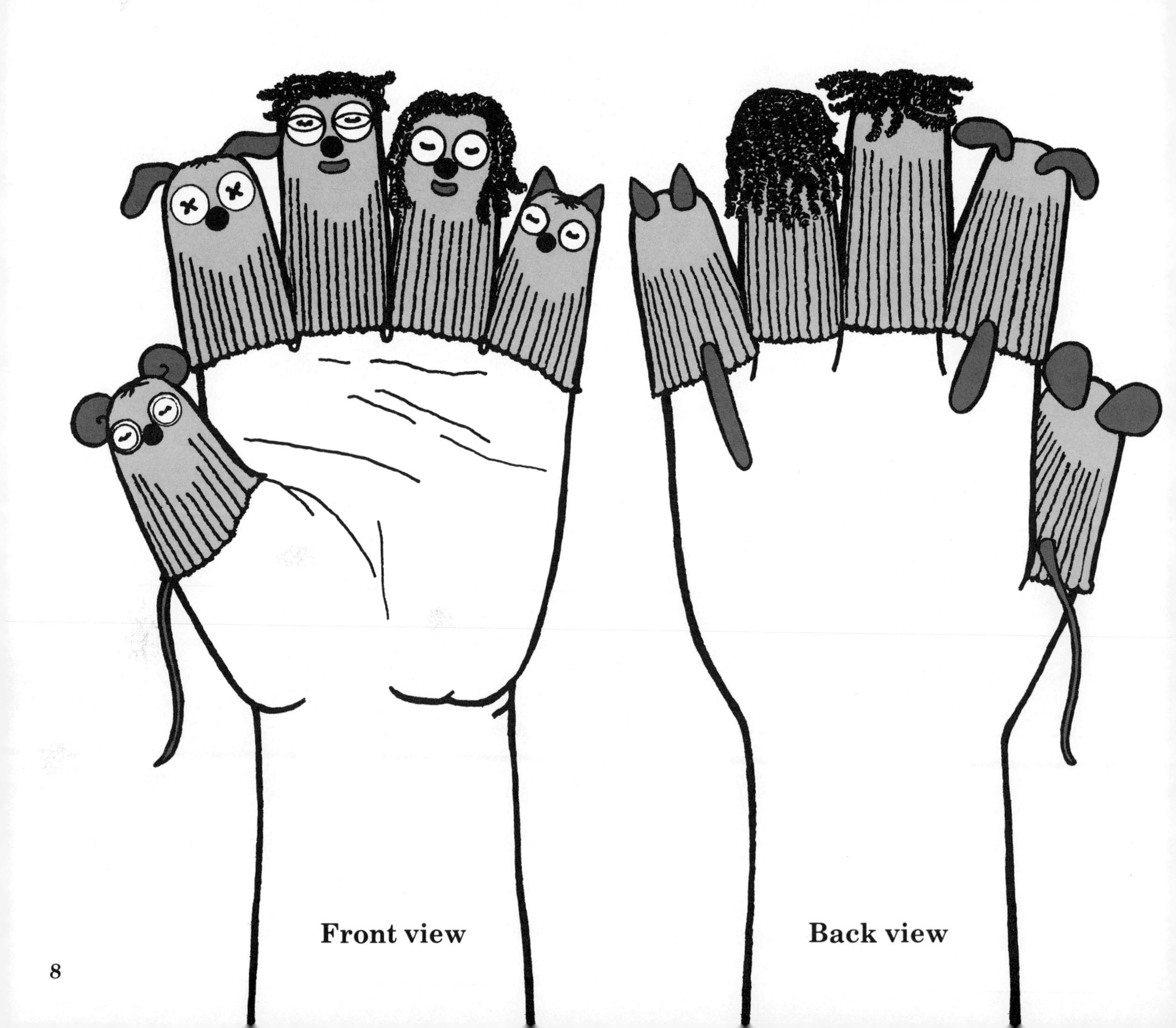
Front view
Back view

PUPPET HEAD

(made from rest of glove)

Sew glove together where fingers were cut off.

Sew on button eyes.

Cut out cloth shapes for a nose and a mouth.

Glue mouth and nose in place.

Sew on yarn hair.
Hi!
Put 3 middle fingers in glove.
The thumb and small finger are
puppet's hands.

GLOVE MONSTER

An old glove and some yarn.

Sew a bunch of yarn on the back of the glove.

Do not sew the glove together.

Draw 2 eyes.

Cut them out.

Glue eyes on yarn.

Put hand in glove. Place tips of fingers on table.
By moving fingers, this monster can walk.

GLOVE DRAGON

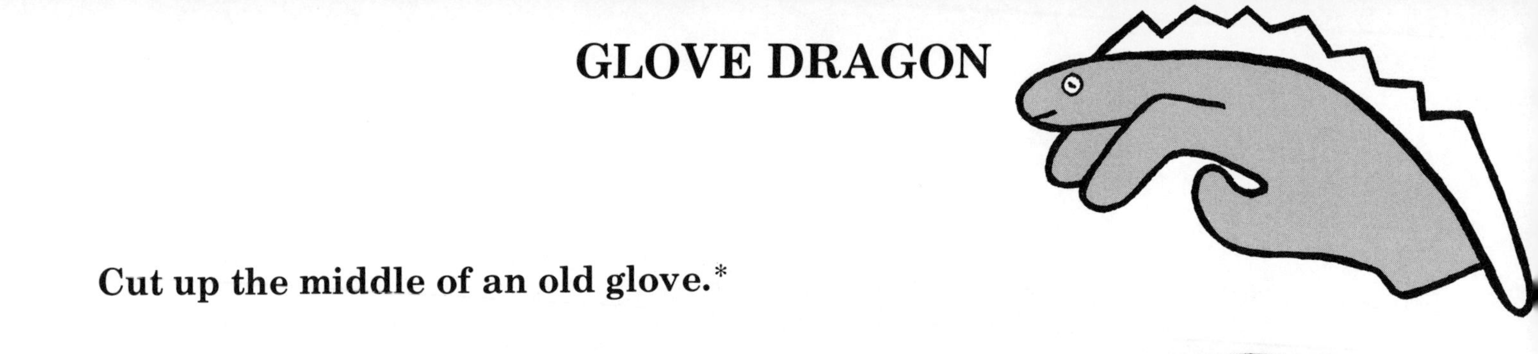

Cut up the middle of an old glove.*

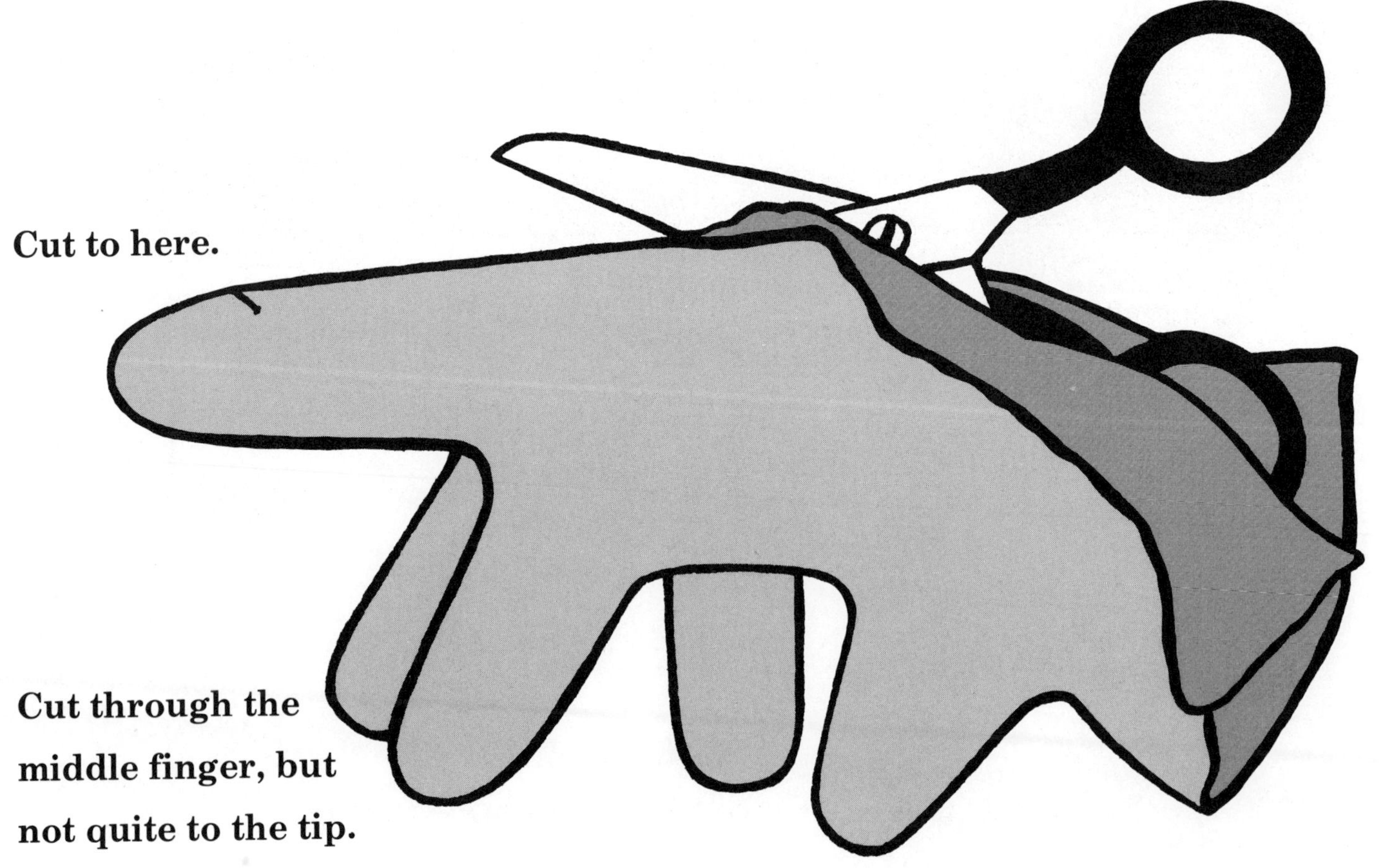

Cut to here.

Cut through the middle finger, but not quite to the tip.

*Rubber gloves make good dragons.

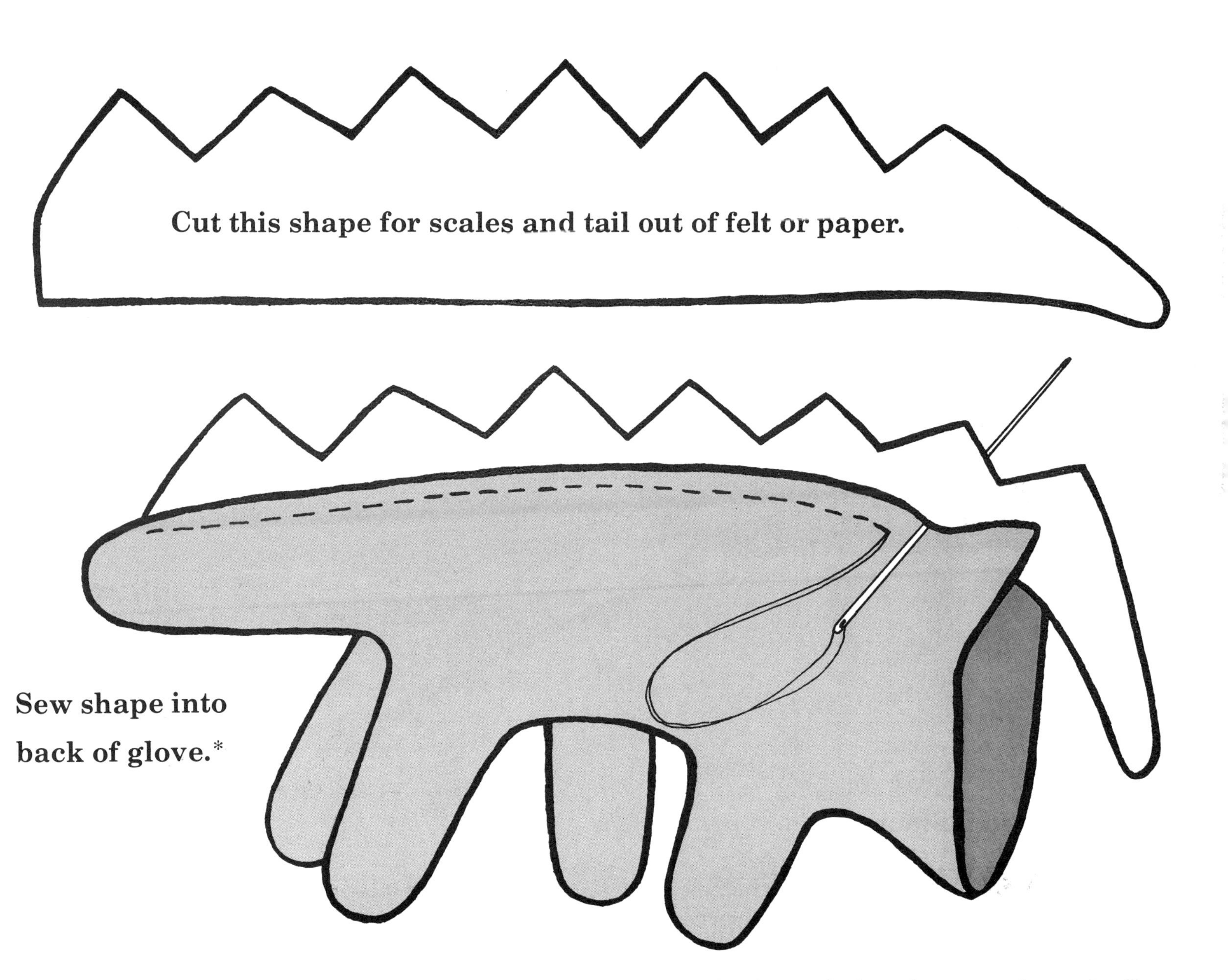

Cut this shape for scales and tail out of felt or paper.
Sew shape into back of glove.*

*If using rubber glove, use glue instead of sewing on scales and tail.

A mouth can be slit with scissors.

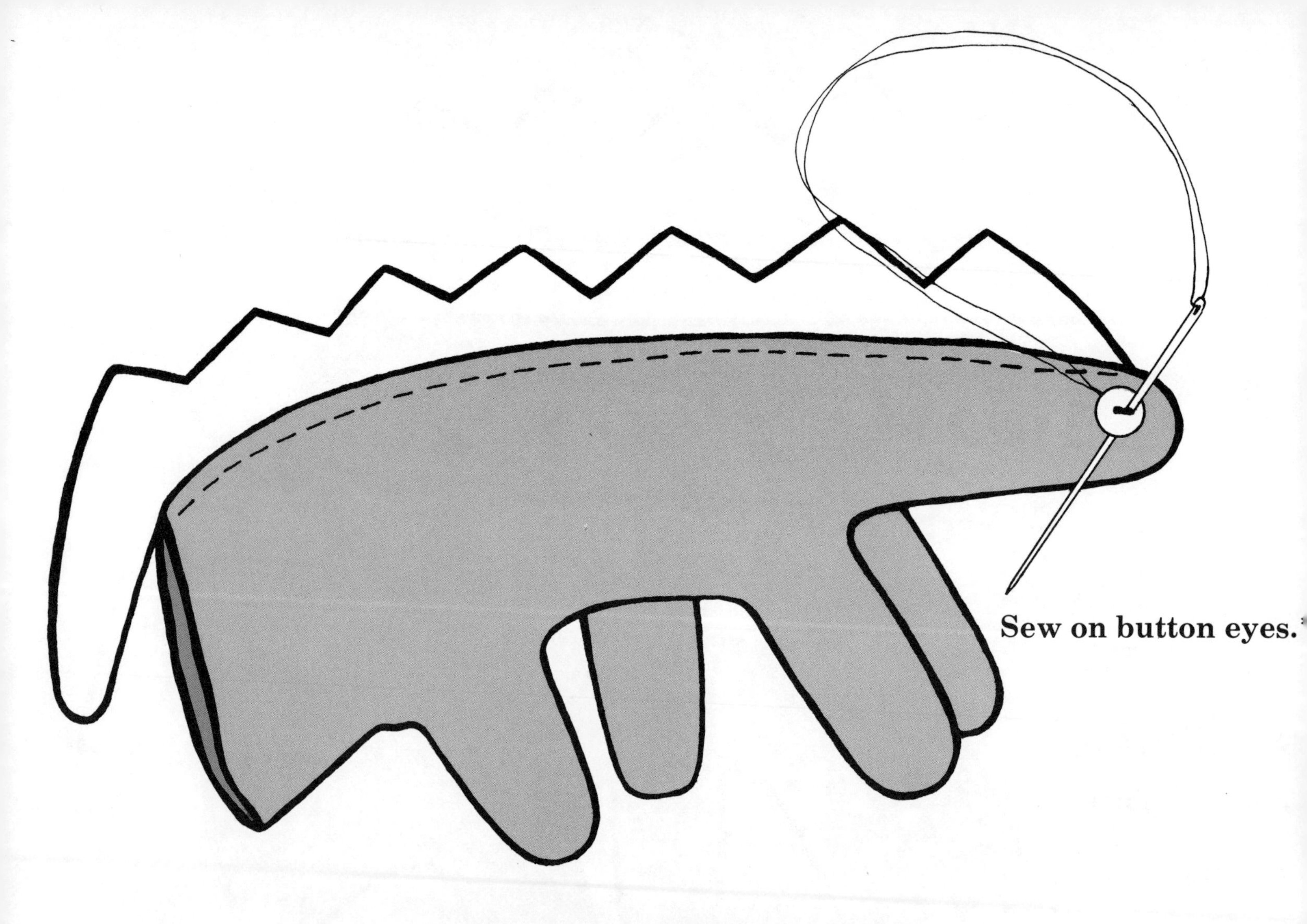

Put hand in glove. Place finger tips on table.

Move fingers, and dragon will walk.

*Eyes can be drawn on a rubber glove.

GLOVE RABBIT

(made from two gloves)

Turn an old glove inside out.
Cut off index and smallest finger.
(Save these fingers to use for feet.)
Sew up openings.

Turn glove right side out, but leave thumb turned in. This is the rabbit's mouth.

Sew on button eyes and a felt nose. Use extra heavy thread to make whiskers.

This is the rabbit's head. The second glove will be the rabbit's body.

Sew the two fingers cut from the head glove onto the bottom of the second glove.

Place glove on hand.
Put three fingers
into rabbit's head.
Now hop along.

SILLY ANIMAL

Turn glove inside out.
Push two center fingers inside glove.
Fingers left out are two ears
and a nose.
Sew on button eyes.

This silly animal head can be
used with the rabbit's body.

MITTEN DUCK

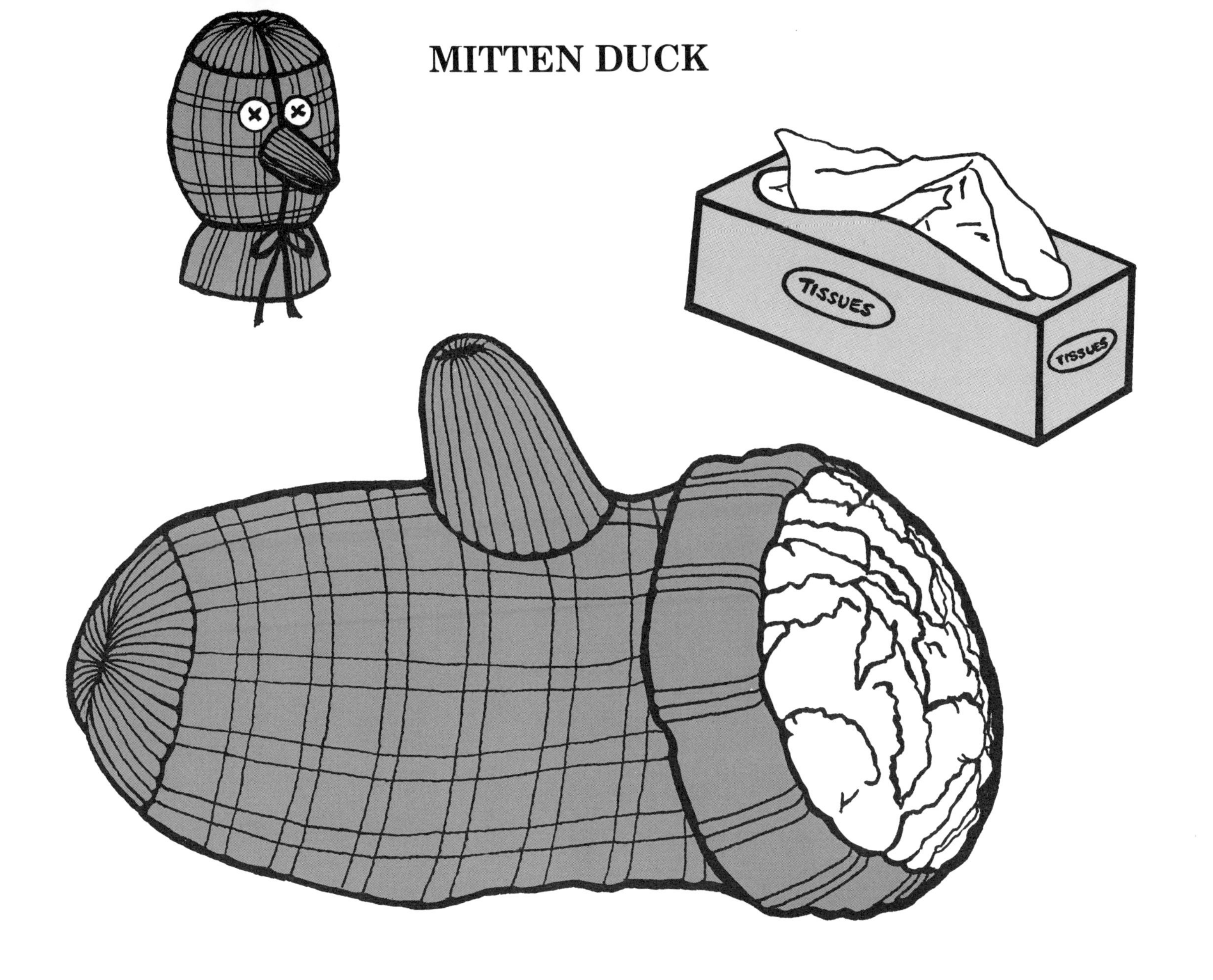

Stuff a mitten with tissues or rags or any stuffing material.

Turn mitten so thumb becomes a beak.

Sew on button eyes.

Tie a string or ribbon around duck's neck.
Put index finger into head.
Make a fist with other fingers.

MITTEN MAN

Cut off the thumb and make a hole in the opposite side of an old mitten.

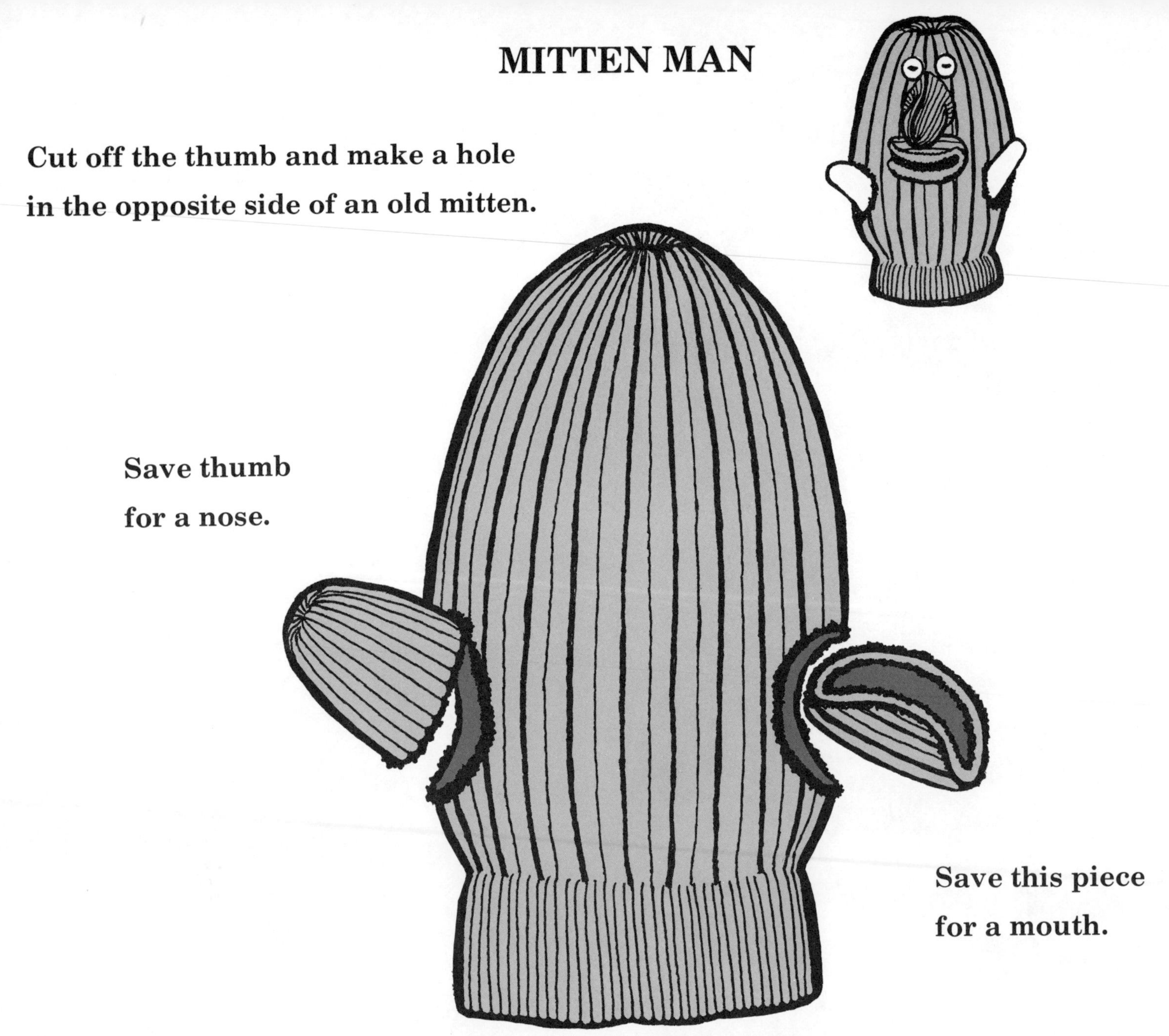

Save thumb for a nose.

Save this piece for a mouth.

Stuff thumb with a cotton ball.

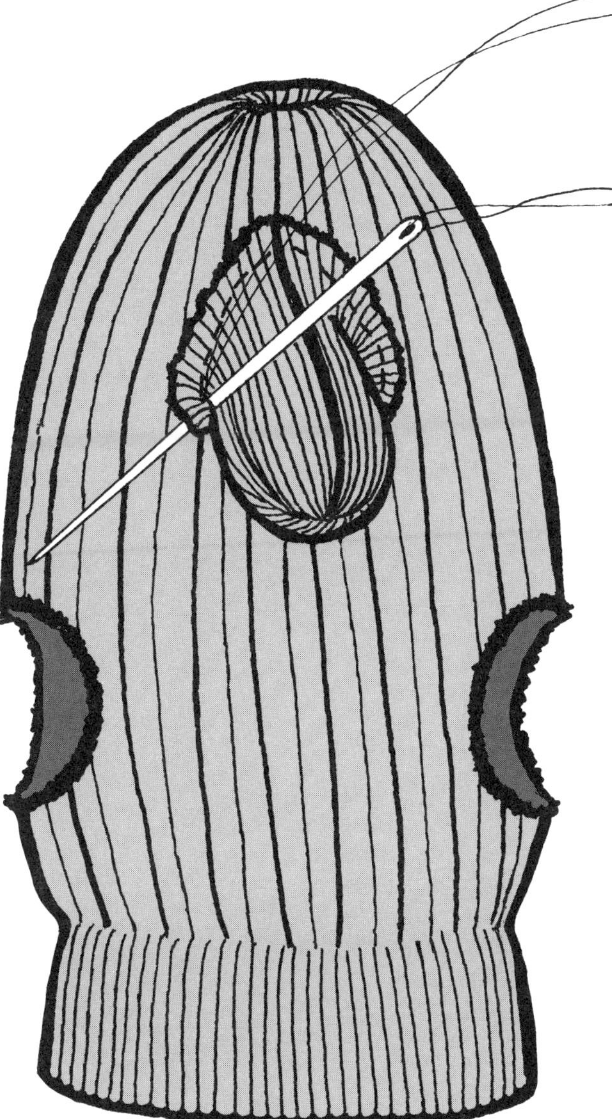

Sew stuffed thumb on front of mitten for a nose.

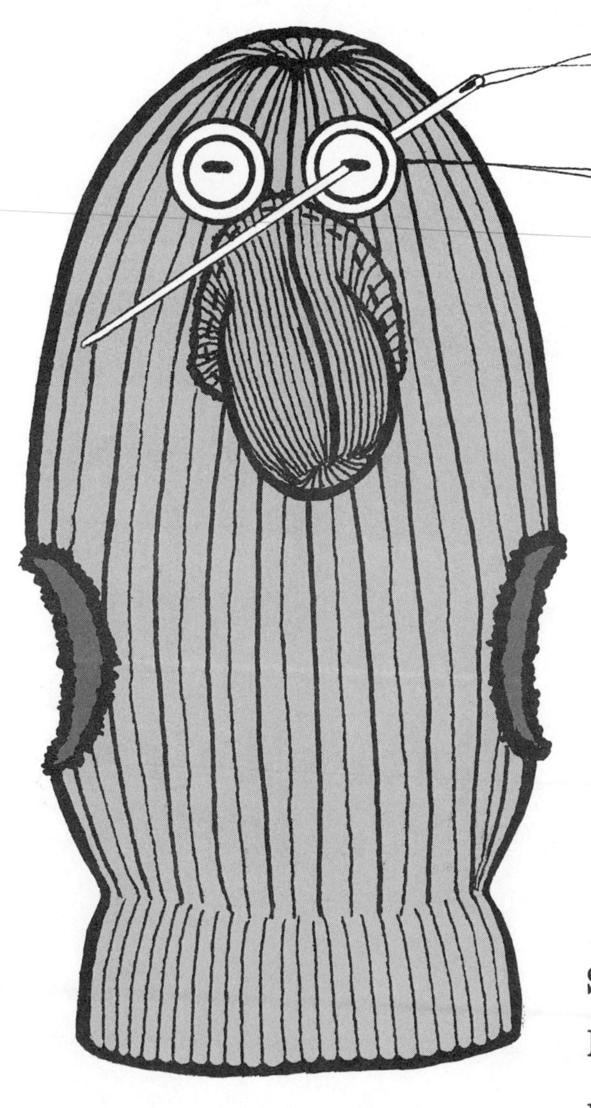

Sew on button eyes.

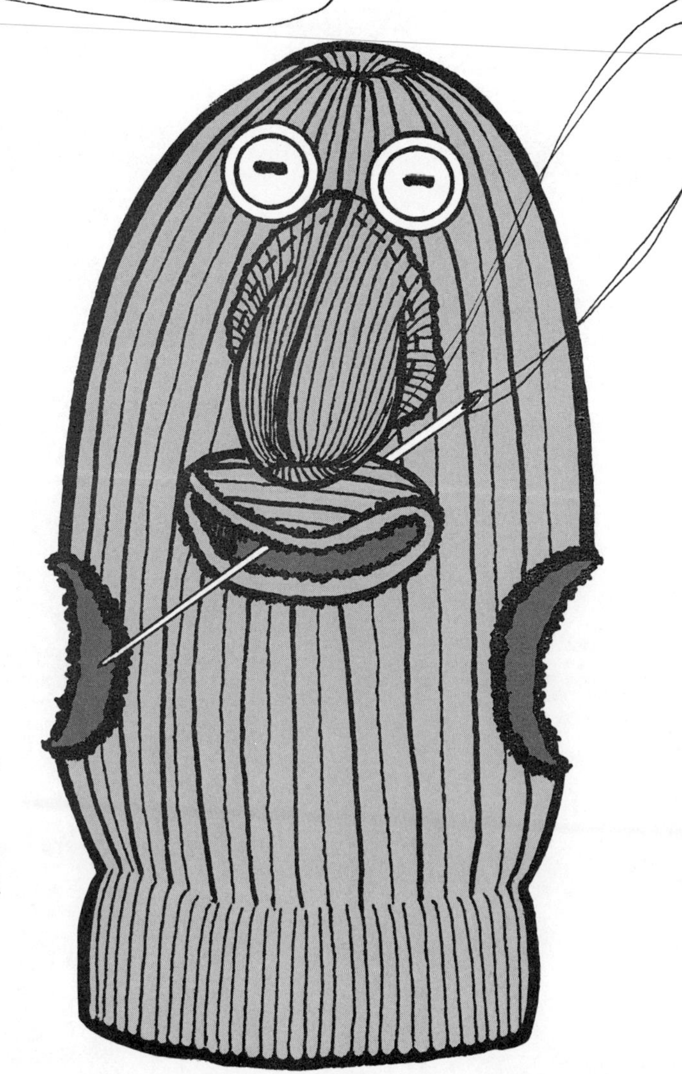

Sew on mouth. Mouth is the piece of mitten cut from side.

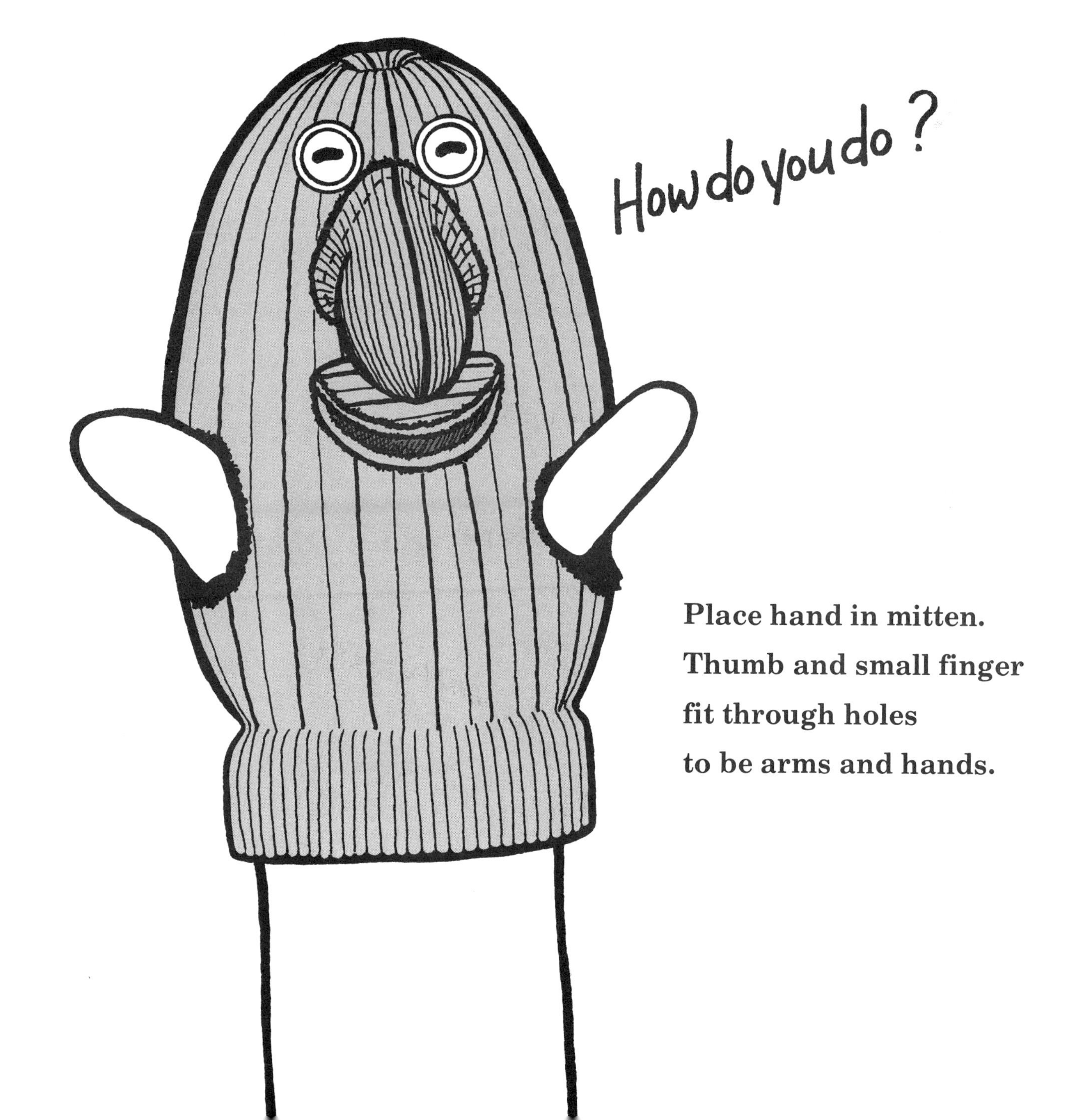

Place hand in mitten.
Thumb and small finger
fit through holes
to be arms and hands.

MITTEN MUTT

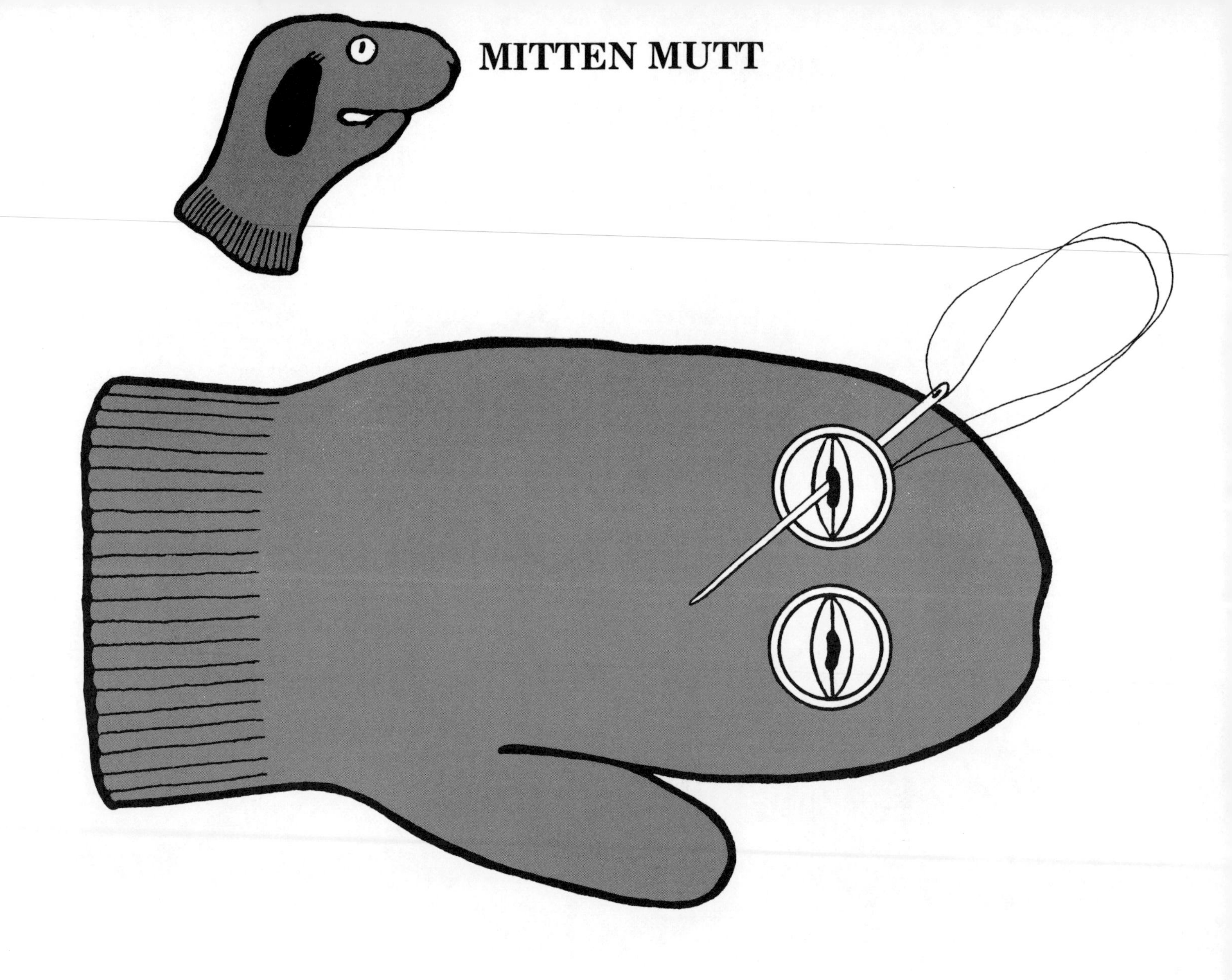

Sew button eyes onto an old mitten.

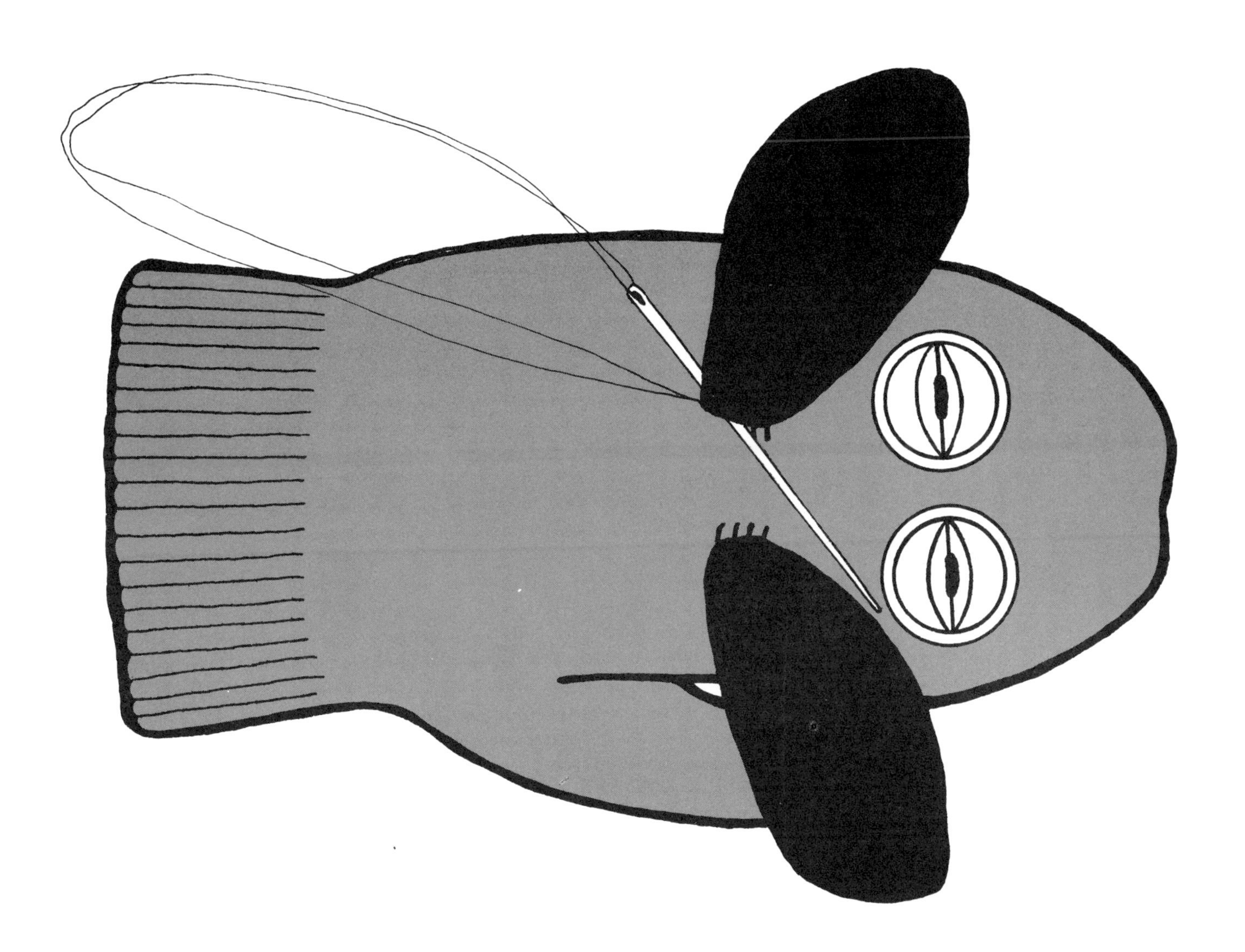

Cut out ears from felt or other cloth.

Sew ears on mitten mutt.

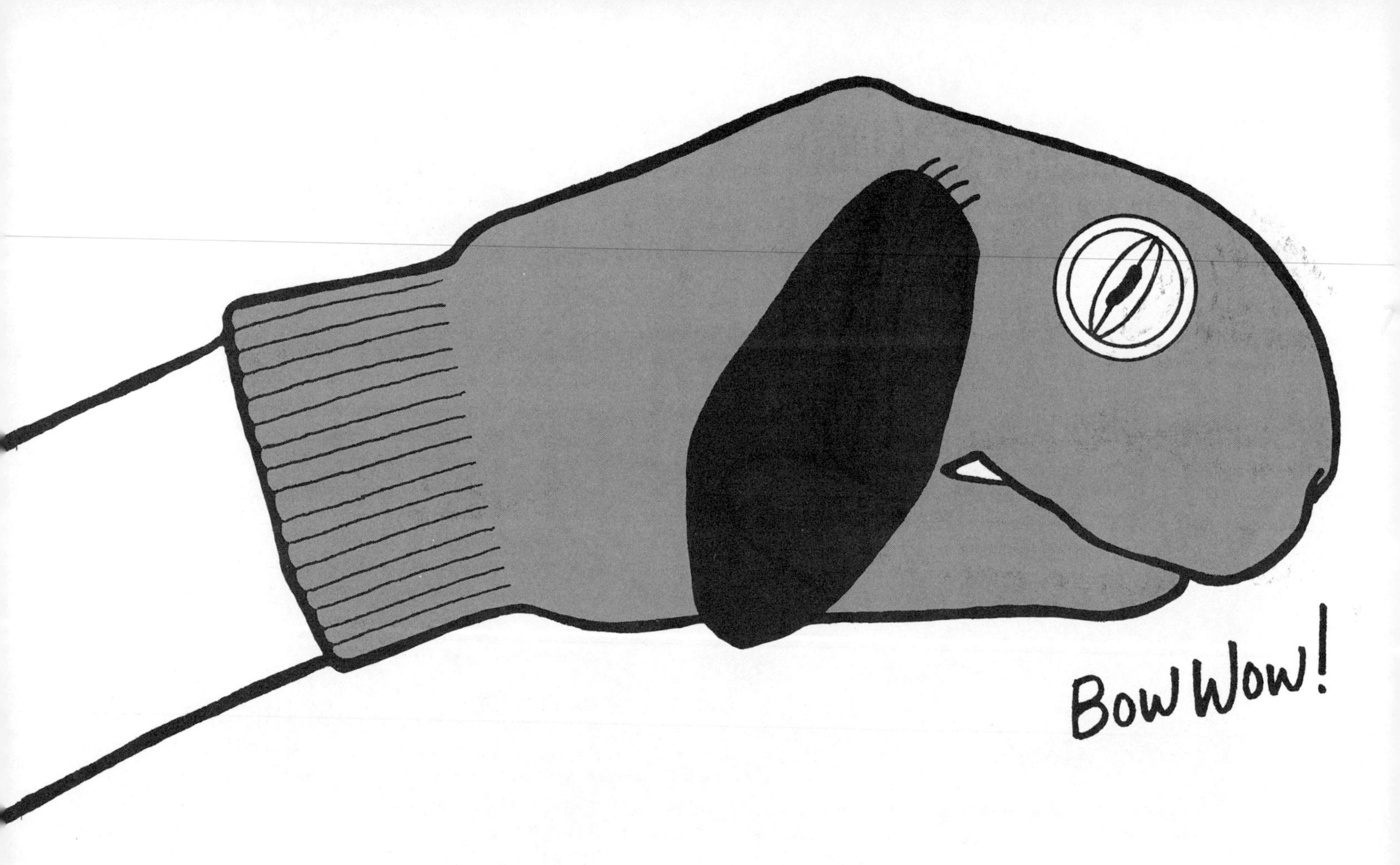

Place hand in mitten.
Thumb is lower jaw of mouth.
By moving thumb up and down,
it will look as if this
mutt can bark or talk.

SOCK SOMEONE

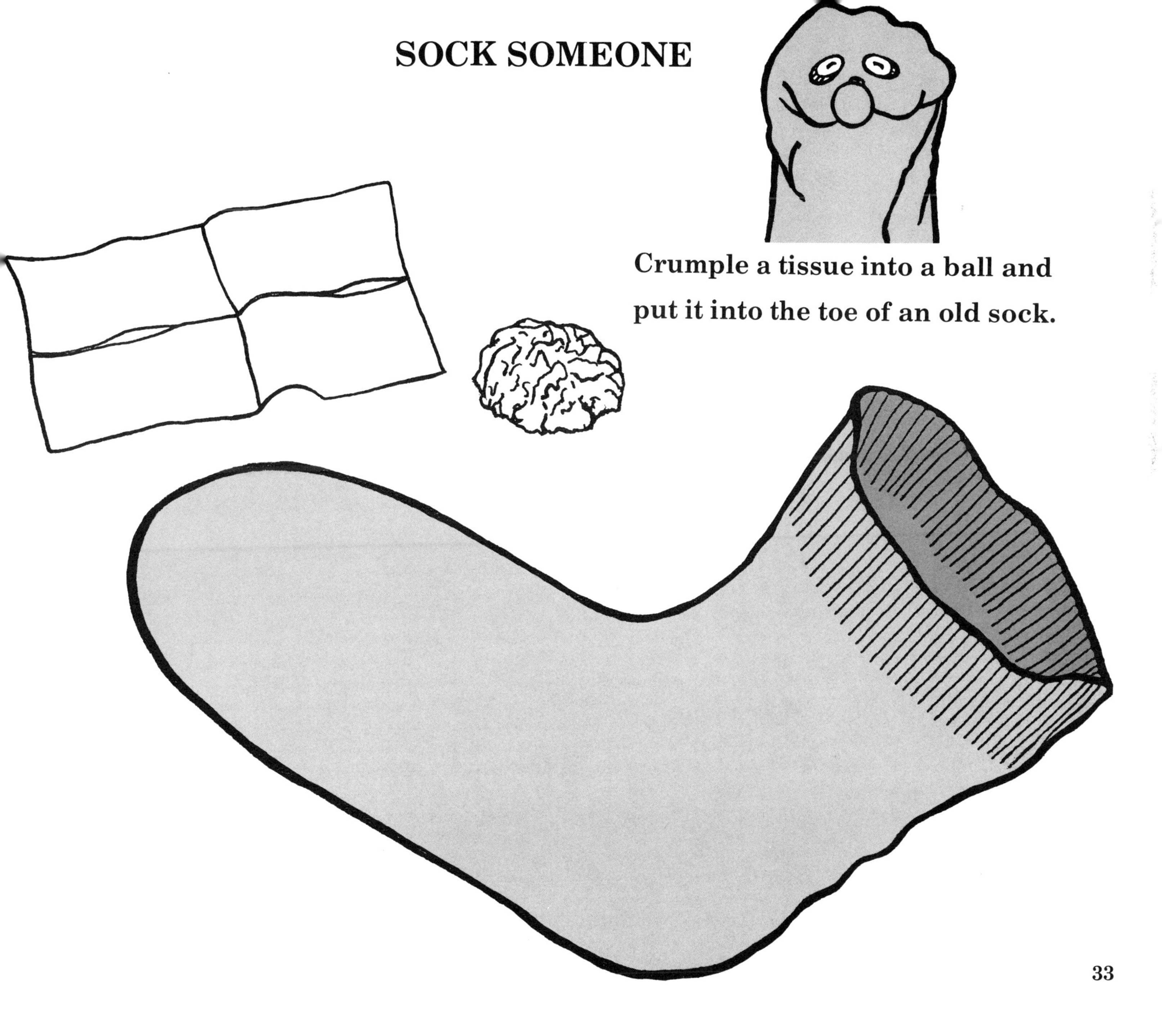

Crumple a tissue into a ball and put it into the toe of an old sock.

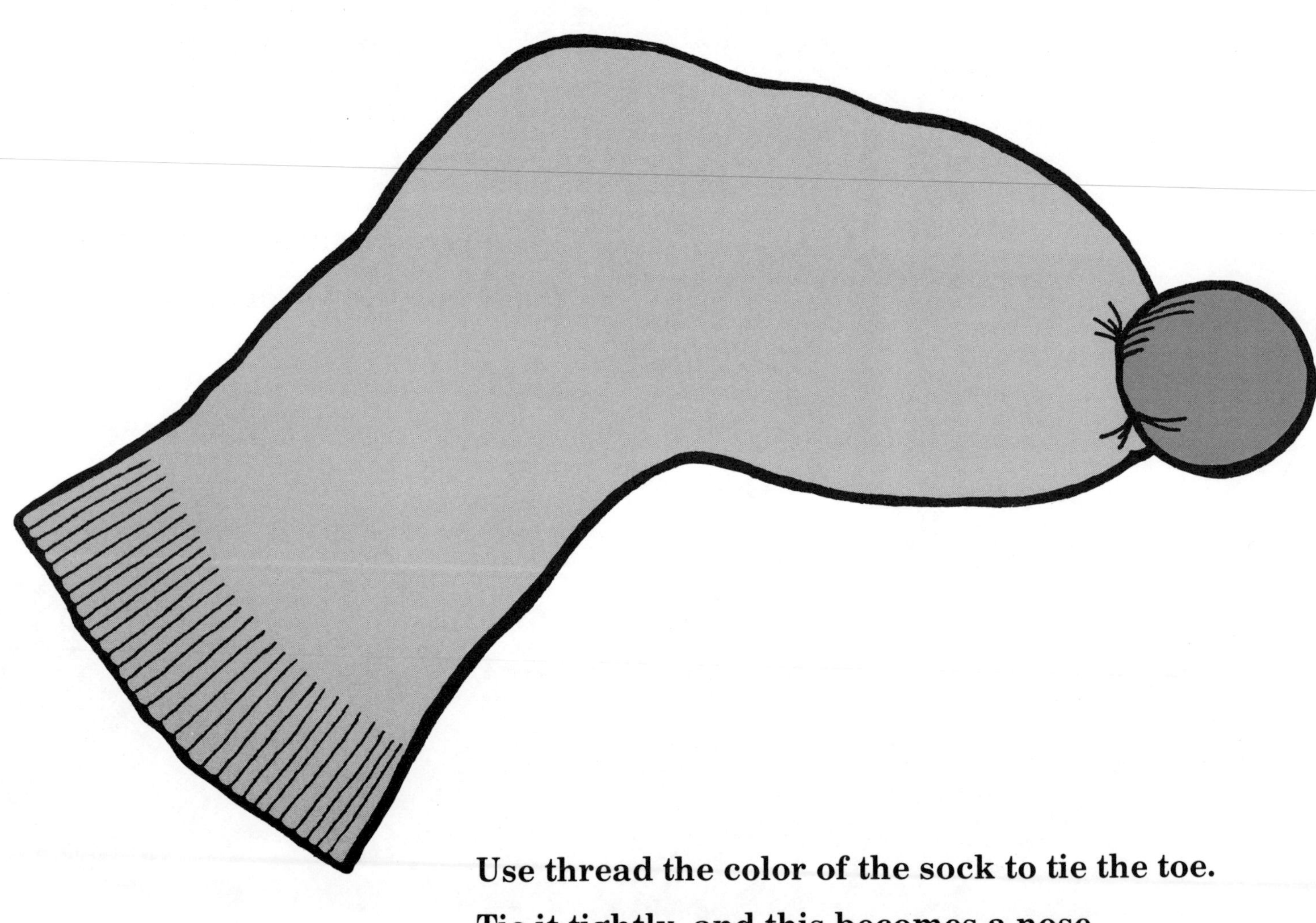

Use thread the color of the sock to tie the toe.
Tie it tightly, and this becomes a nose.

Sew on button eyes.

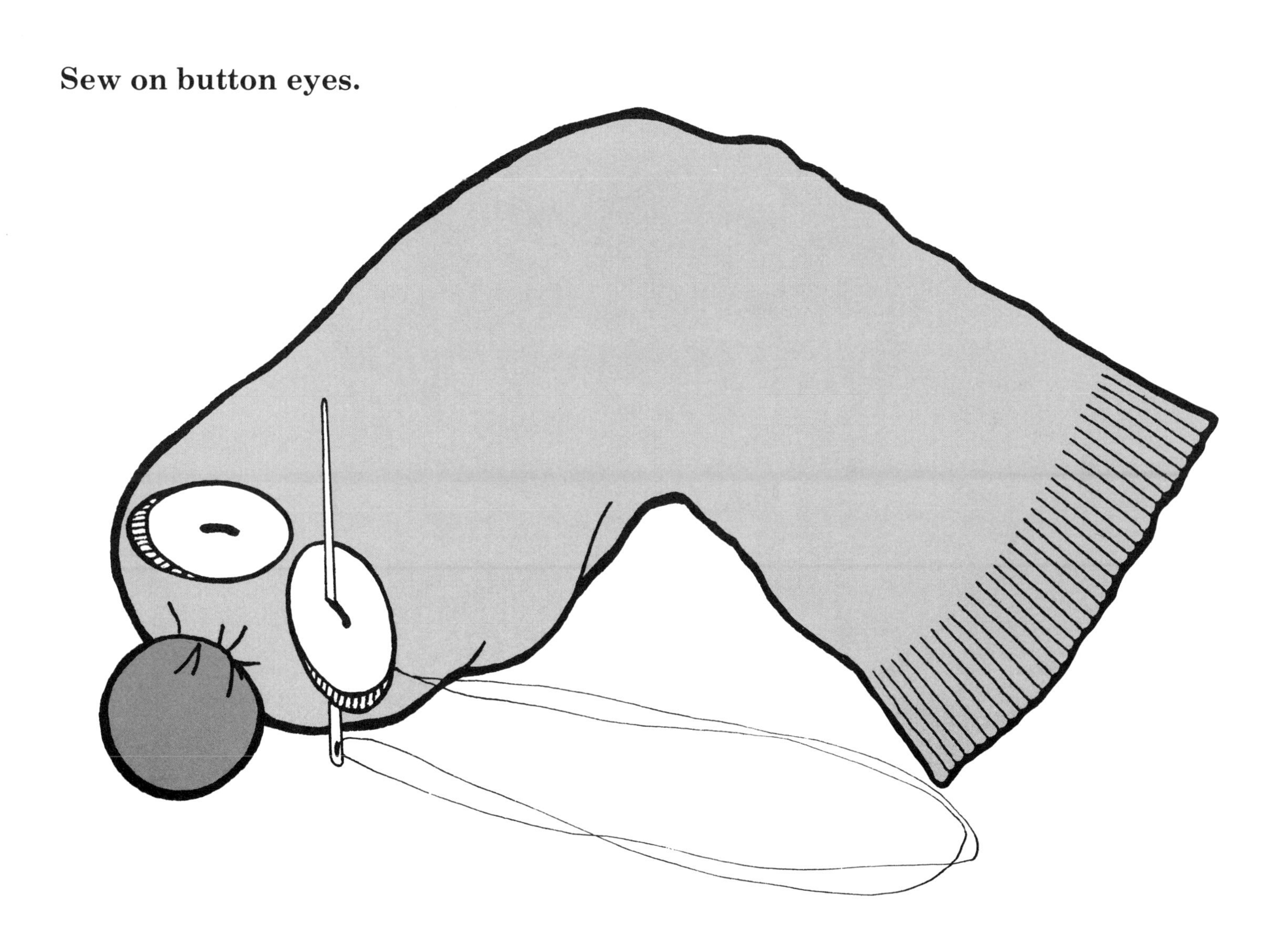

Put hand in sock. Thumb will be lower part of mouth.

This puppet can open and close its mouth.

SOCK MARTIAN

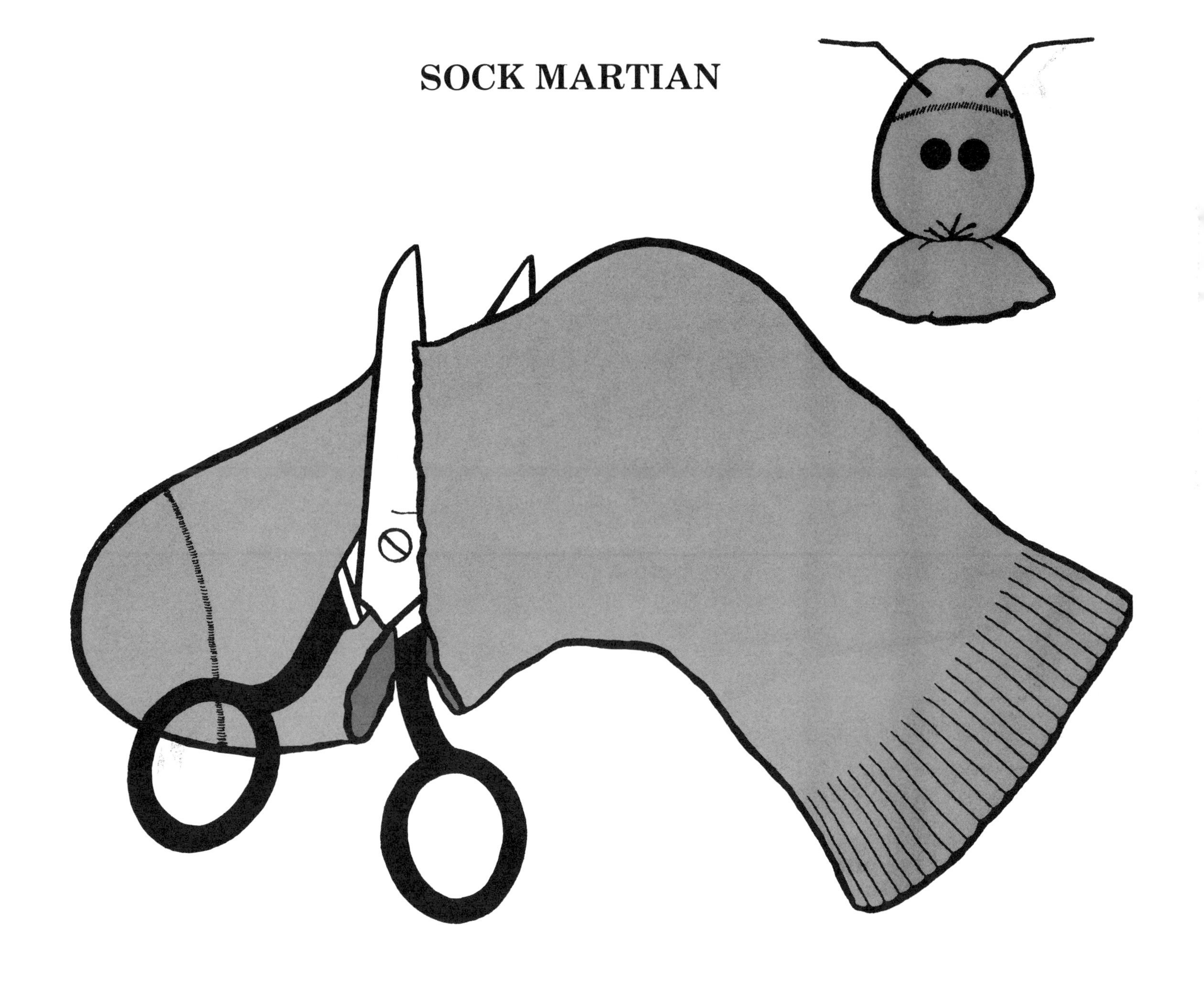

Cut off half of an old sock between the toe and heel.

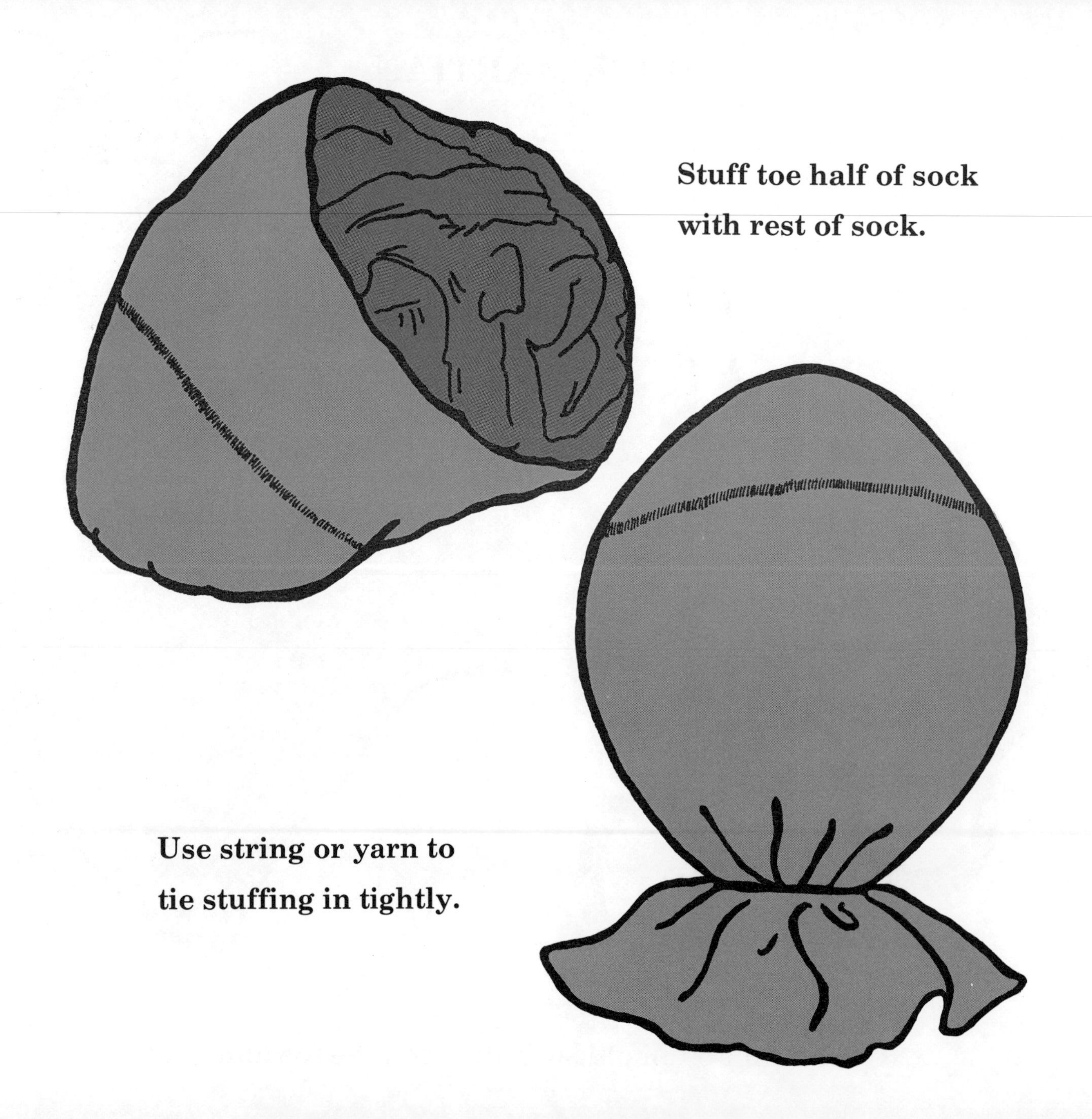

Stuff toe half of sock with rest of sock.

Use string or yarn to tie stuffing in tightly.

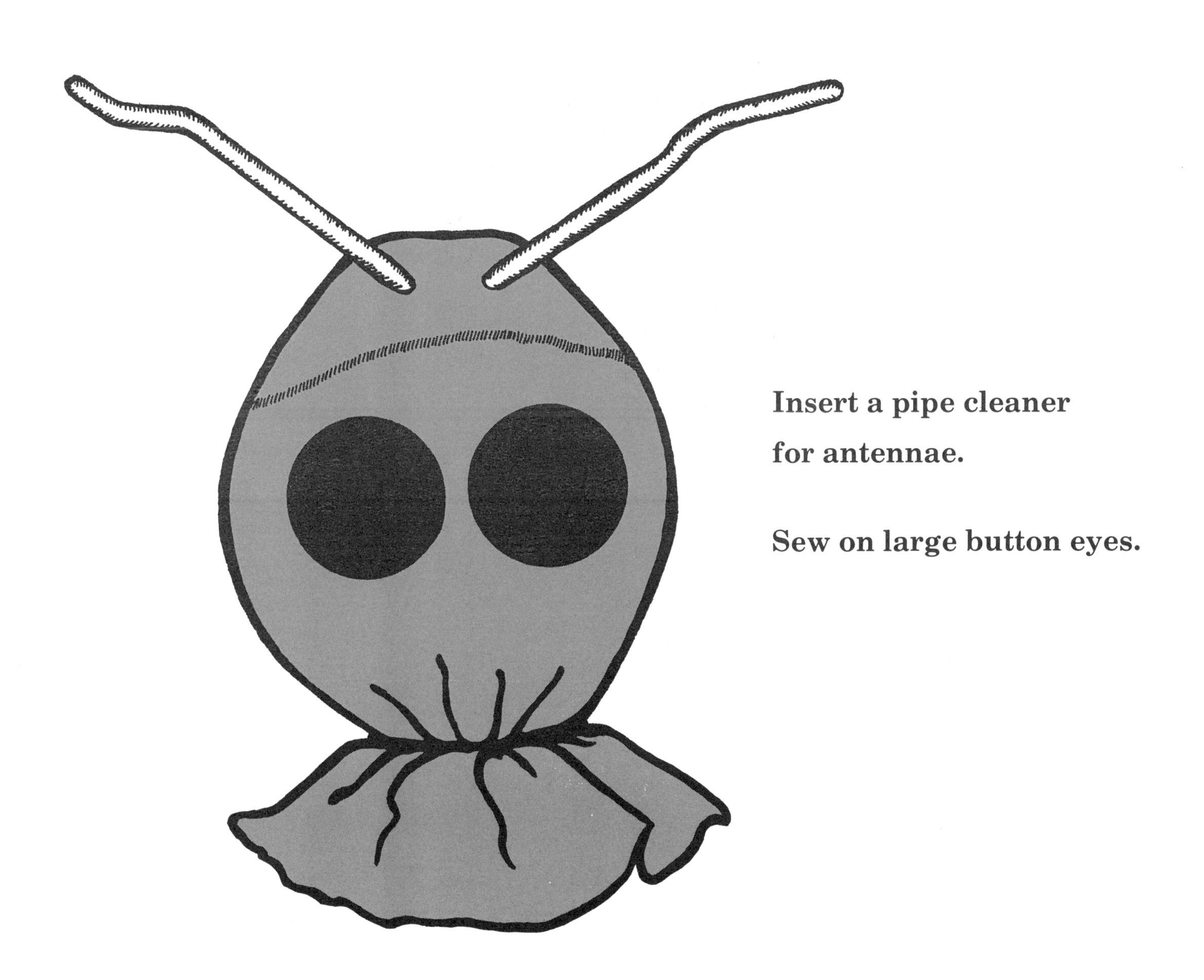

Insert a pipe cleaner for antennae.

Sew on large button eyes.

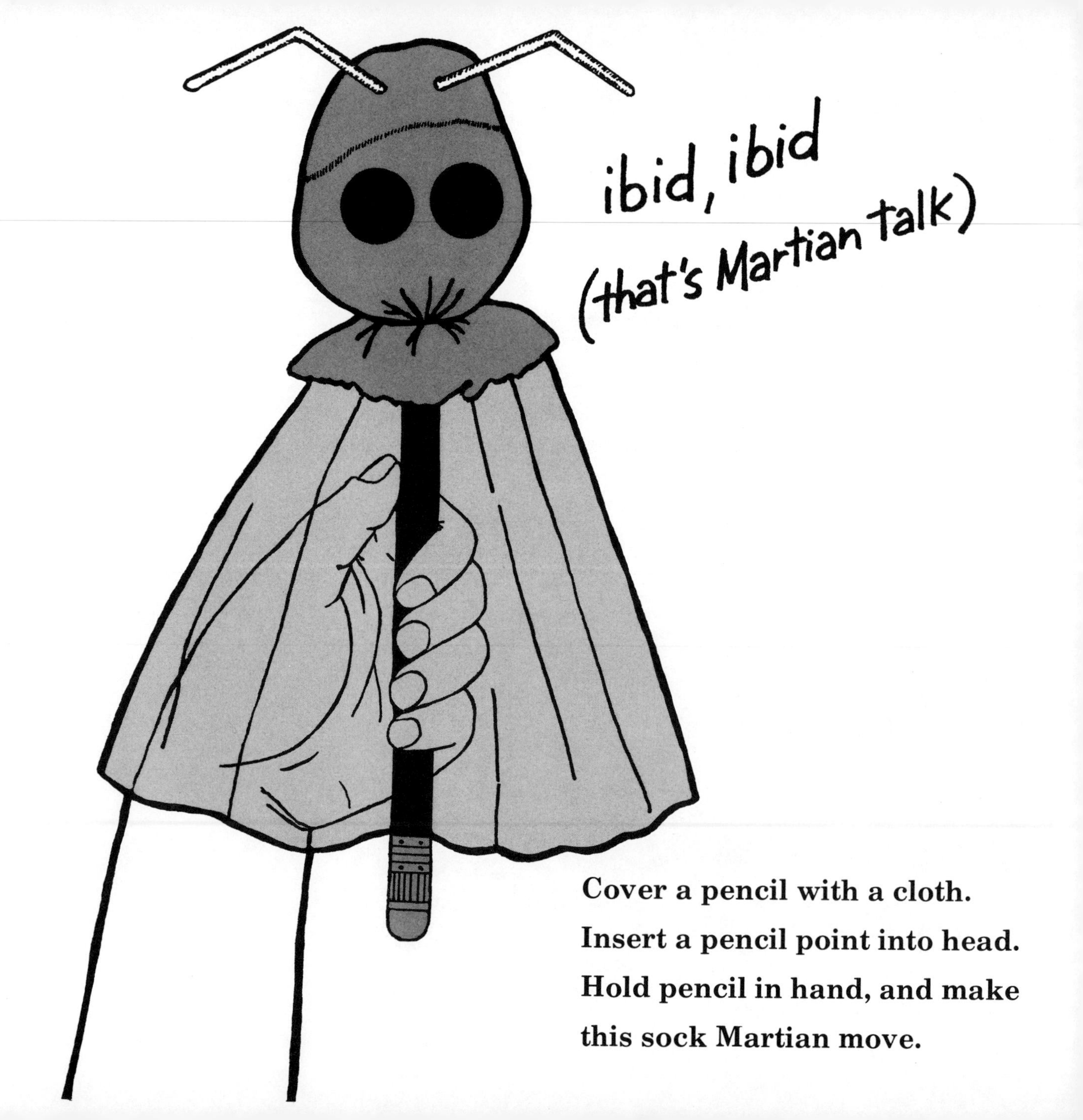

Cover a pencil with a cloth.
Insert a pencil point into head.
Hold pencil in hand, and make
this sock Martian move.

SOCK HORSE

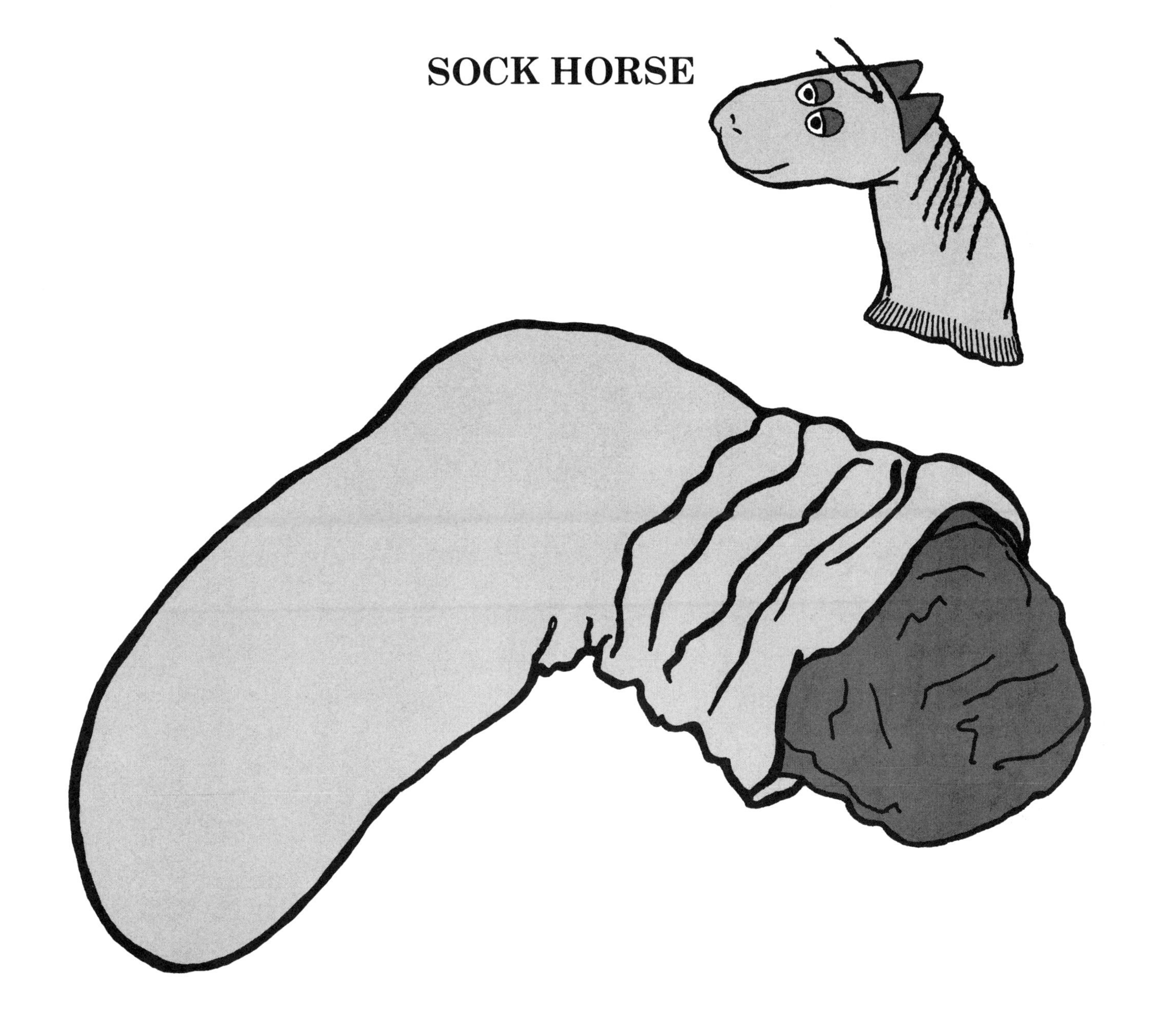

Use tissues or rags to stuff the foot of an old sock.

Draw eyes on paper. Cut them out and glue onto face.

Draw nostrils and a mouth with a felt tip pen.

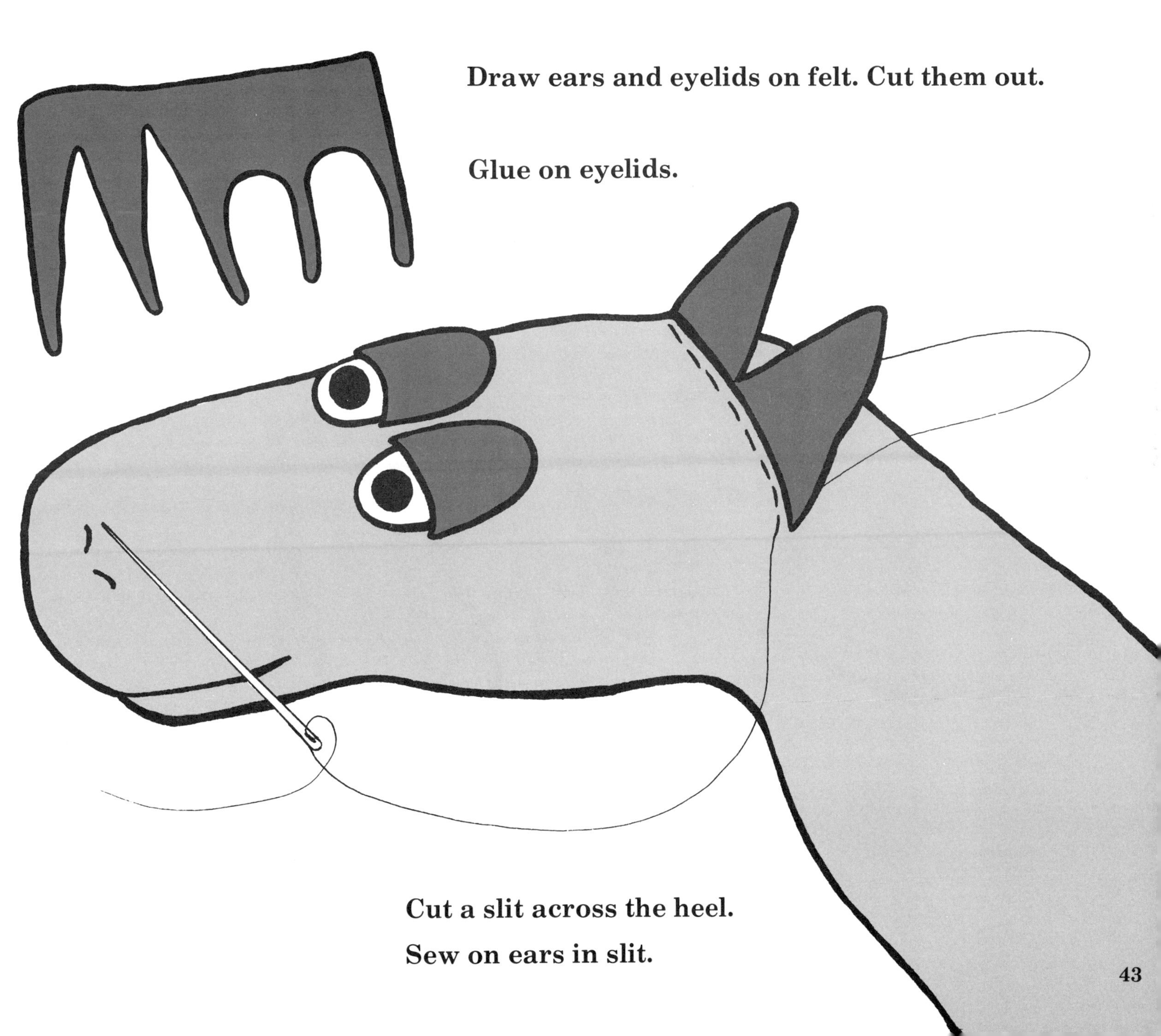

Draw ears and eyelids on felt. Cut them out.

Glue on eyelids.

Cut a slit across the heel.

Sew on ears in slit.

Cut four 8-inch lengths of yarn.

Make one set of holes in front of ears, and three sets behind ears. Pull yarn through holes. Tie a knot. The yarn will be the horse's mane.

KNEE SOCK SNAKE

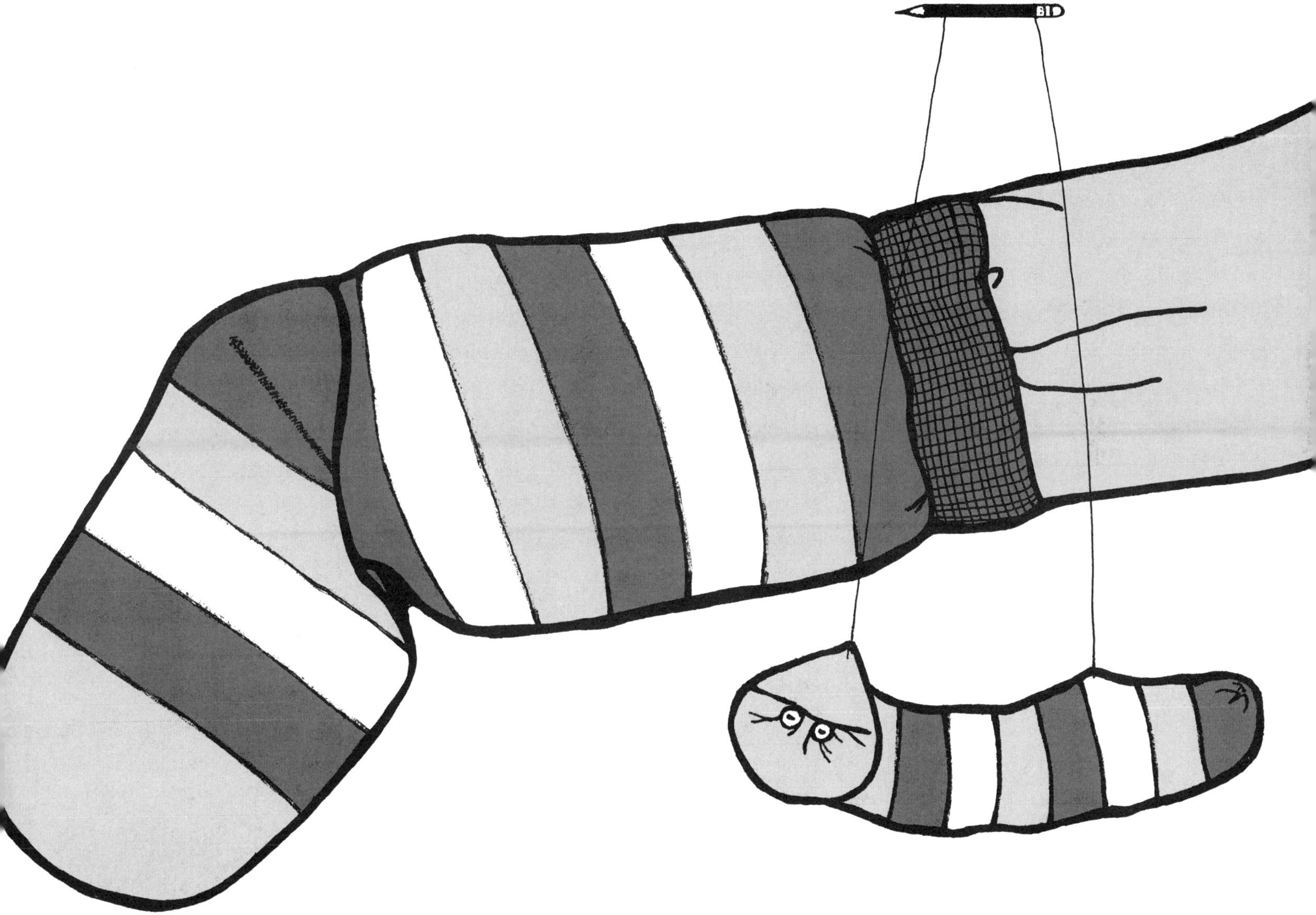

Stuff a knee sock with a rag rolled into a tube shape.

Pull a thread from the top of the snake's head.

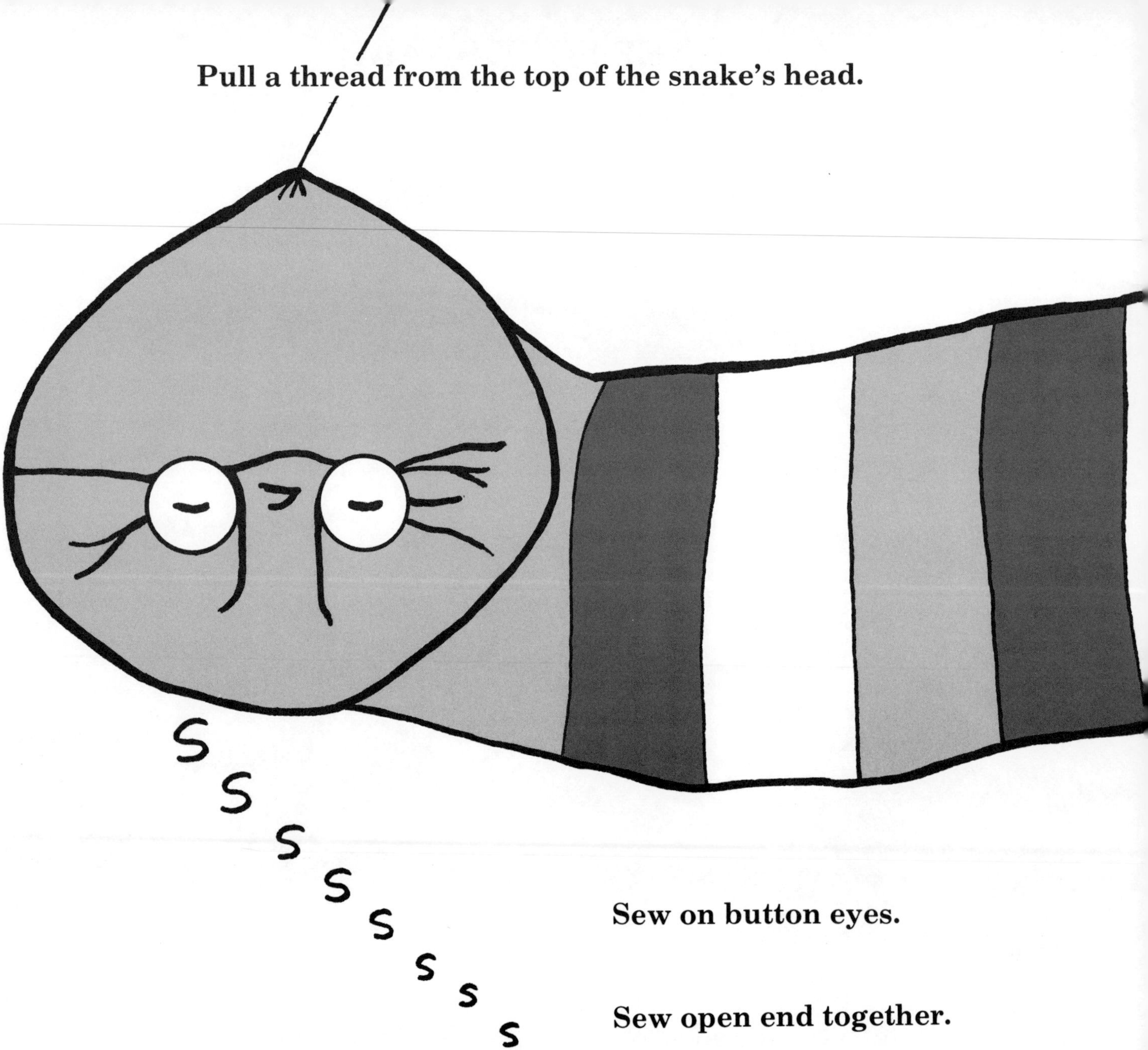

Sew on button eyes.

Sew open end together.

Pull another thread from the middle of its back.

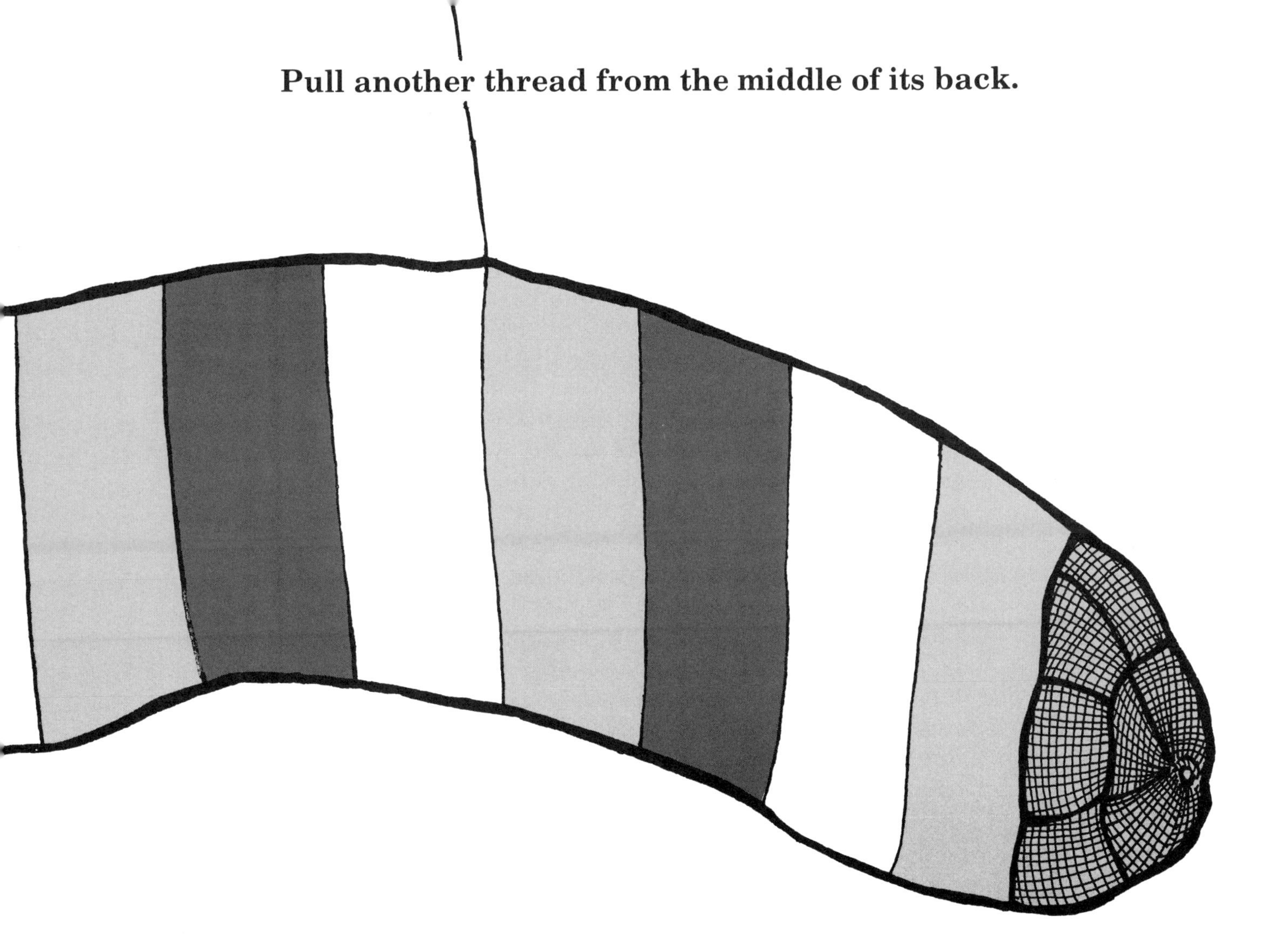

Tie thread ends to a pencil.

By moving pencil, the sock snake can crawl and move its head.

Puppets have played a major role in the life of author/illustrator Frieda Gates. As a professional puppeteer, she performed with the Hudson Valley Vagabond Players. She is a member of the Puppetry Guild of Greater New York and of The Puppeteers of America. Mrs. Gates studied art at the Brooklyn Museum Art School, The Art Students' League, The New School for Social Research and New York University. Currently she teaches design, drawing, painting and production at Rockland Community College in Rockland County, N.Y., where she makes her home with her husband and their three children. Mrs. Gates' paintings have been exhibited in New York City, Rockland and Westchester counties, and are contained in private collections.